HE IS NOT MY ANCESTOR

by

Ahmed Ismail

TRAFFORD
PUBLISHING

Note for Librarians: A cataloguing record for this book is available from Library and Archives Canada at www.collectionscanada.ca/amicus/index–e.html

ISBN 1-4251–0791–5

PUBLISHING™

Offices in Canada, USA, Ireland and UK

Book sales for North America and international:
Trafford Publishing, 6E–2333 Government St.,
Victoria, BC V8T 4P4 CANADA
phone 250 383 6864 (toll-free 1 888 232 4444)
fax 250 383 6804; email to orders@trafford.com
Book sales in Europe:
Trafford Publishing (UK) Limited, 9 Park End Street, 2nd Floor
Oxford, UK OX1 1HH UNITED KINGDOM
phone +44 (0)1865 722 113 (local rate 0845 230 9601)
facsimile +44 (0)1865 722 868; info.uk@trafford.com
Order online at:
trafford.com/06–2549

10 9 8 7 6 5 4 3

✳

CONTENTS

INTRODUCTION

Knowledge is a wonderful thing/dangerous thing. It is a double-edged sword capable of healing and wounding. It can open up the mind as well as polluting it. Without knowledge and the imparting of the same, mankind could revert to a mindless creature.

At various times and places man has used knowledge to subdue his planet, bring forth an understanding of what is called the 'laws of nature', develop inventions to make life easier and has allowed himself to express law and order on cultures. In fact, knowledge has been used to pass on the various religions and philosophies now current in the world.

On the other hand, knowledge has been used to create chaos and disorder, to abuse the laws of nature and to destroy parts of the human race as well as to subvert the normal thinking process.

It seems that knowledge has two faces: one for the good and one for the bad. That is what it seems but in fact knowledge is a servant of man and as it turns out man is the one who can boast of having two faces. Therefore, knowledge is like fire. Fire helps man to make iron, cook his food, warm his house and even send man into space. It also has been used to purposefully kill people, generate illegal money as in arson and destroy unwanted objects.

With this understanding, knowledge is useful only when it is considered practical, helpful and enlightening and when it can reflect the relationship between God and man. The idea of the relationship between God and man bothers some people, who for lack of a better term are not knowledgeable. They would consider the term restrictive whereas, in truth, the term between God and man means between himself and God, himself and other human beings, himself and nature, himself and his beginnings and his final return to an everlasting life and his acts of what is called worship. Other than that, knowledge is worthless.

In this materialistic world, man has to cash-in on the value of knowledge to avoid carrying around useless stuff as excess baggage. The inventions of the phone, television, computer and the car have been used for beneficial purposes to spread kindness, love and offer assistance even in the help of worship. These inventions have also been used to defraud, kill, maim and cause irritation. So the end result is how one can use knowledge for beneficial purposes rather than destructive ones, which in itself is a form of worship.

The highest form of knowledge is the knowledge that even though it may hurt to hear or learn about it, would set one free of falsehoods that hinder men from the truth. That sounds so blasé in this modern world but it is absolutely true. This idea goes back to the brotherhood of prophets and to why they came and what they came to impart. For even in this sometimes dull, thoughtless world, people still realize that the sanctity of the office of prophethood was instituted to bring advice, guidance, counsel and knowledge to people. Knowledge was imparted to lift mankind up to a dynamic creature (vicegerent) rather than abase him to below the level of an animal.

If knowledge is used to free people from misconceived practices or beliefs, then knowledge is good. Man is a curious creature but he also has tendencies of being selfish such that if he were to get one mountain of gold, he would desire to possess another mountain of gold even if his lust were to degrade others in his quest. Therefore, man is not left all alone to wander here and there but is given practical rules to live by. What he does with these rules is his own business but he has been told that he will be accountable for his actions on the Last Day of Judgment.

This should have a horrific impact on his thinking because of the awe it should demand on his actions but especially in these times it seems like most of us prefer common nonsense to common sense. Worldly smarts is not the real answer. In fact, worldly smarts tends to lull man into a state of drowsy sickness that like any disease is actually spreading. One of the symptoms of this disease is a false sense of accomplishment that puts the idol of worldly fun and adventure in the forefront of life.

Nevertheless, if life is precious and not just a toy-thing, then the life in the Hereafter is even more desirable – much more desirable. So in this respect, mankind chooses his own road and should spend some time to explore the highway of life that he is on and to do so with a form of earnest mentality. If he is going to be judged, he should know as much as he can about the criterion for his judgment. Or like a bombastic ass, he can choose to throw it all up in the air and let what he might call 'lady luck' take care of things. At least he should have the intelligence to choose correctly and the courage to investigate his choices. Only a fool or an idiot would be either arrogant or ignorant enough to consider himself self-sufficient. (See appendix A for a discussion concerning knowledge).

A person should have at least a modicum of appreciation for the skills shown by the valiant men and women who endeavor to make sense (and thereby bring some knowledge to us) out of this world. Most of these people we call scientists. The true scientist is the one who tries to make sense out of what he can fathom and bring the world into a kind of sensible state of affairs. However, in reality, knowledge comes in drips and drabs and theories are advanced and put on the junk heap when more refined knowledge becomes available. Basically, this occurs in what can be called the theoretical sciences which for the most part deal with philosophical-religious things. Even physics, chemistry and biology are not exempt from this as far as they go but they are on a more solid ground due to their practical, observable natures.

The science of paleoanthropology is a science like many theoretical sciences which depend on money, prestige and the ability to excite and titillate. It is also a science that tends to go into dangerous zones of human or in this case pre-human activity and make its philosophy more concrete and palatable than it really is. It is also burdened by a western trend of divorcing God and man.

To fix a problem one has to see the problem and admit some fundamental, basic things. What is wrong in that? So paleoanthropology is fundamentally flawed but not necessarily incorrect. So what is wrong in that? To err is human and anthropalentologists are as human as the next person. It is only natural to do some honest soul-searching especially when caught out and try a more tactful and honest approach.

In this book they will be taken to task and an alternative theory to mankind's 'evolution' will be presented in brief.

This alternative theory will not be able to plug all the holes and answer all the questions but neither can the paleoanthropologists. The big difference in the approaches lies in the evidence and type of evidence that can be presented. The evidence for this 'new' theory is not scientifically concrete but it is based on the consistency of pattern development.

The paleoanthropologist can find a bone (concrete) but then has to wrap a story around it (not concrete) and provide a theory. They have provided different theories up till now which have cancelled out other previous theories that they have made.

It would be interesting to put them on truth serum to see how they would react to what they believe in. How refreshing it would be to examine the mental-psychological-historical thinking processes of these scientists as they plow through their logical thoughts and come up with their theoretical positions – from their battle of church versus state, to humanism to getting by on the money issue by avoiding certain 'problems' might make fascinating reading.

Most paleoanthropologists hold to the theory that man evolved from certain apes and or monkeys. There is more to it than that but the gist of the matter is that an animal eventually changed into a man. The way it looks now according to their latest theory is that several animal-types changed into the human race at various times and places to various degrees of success or failure.

Islam rejects the above thinking as to evolution from animal to man as outright false. Islam does not reject that the physical evidence found by these scientists is false. In other words, their theory is based on utter conjecture with no statements that can be sanctioned from the science of religion that by itself deals with the creation. Ignoring that issue puts science on a shaky ground. Admittedly, very few have delved into that science and therefore it is conjectured by some that it has very little to offer in the way of evolution.

Wild conjectures based on non-conducive factors and spurious logic is not the way to go. Besides this, inevitable conclusions by going down a different road have been made in this book showing just how strange the theory of human evolution from an animal to man is. Another damaging thought concerning those who favor animal to human evolution comes from their well recorded history of theory building by which one theory is replaced by another more 'modern' theory which in turn is replaced by a still more up-to-date theory and so forth and so on. This makes their theorizing somewhat spurious and unpractical.

The measure of the man or woman and their science should welcome a logical and practical debate and a voluntary readjustment of values that could be called courageous and honest if the science of paleoanthropology is honest. The result could only be beneficial for everyone concerned.

From prehistoric man to the future of the human race, people have been putting their thoughts in motion. It is very peculiar that a lot of people might be shocked to know that they have many values closely associated with the pagan Arabs of old – especially when it comes to religious thinking. That is a radical thought. When we investigate this possibility, are we really much different than those people?

The pagan Arabs ran into a puzzle that confounded them. His name was Prophet Muhammad (peace and blessings be upon him). These pagans were called non-believers and it shocked them to hear this. The average reader might think that they were called non-believers because they didn't believe in Allah (God) or perhaps they believed in only an Arabic god which they called in their language Allah.

That thinking is absolutely false. The danger for us is how close are we to those pagan Arabs? The pagan Arabs caused a great deal of trouble to Prophet Muhammad (pbh) and his little band of loyal followers for well over a decade. Some of the beliefs the pagan Arabs had are as follows:

1. They believed in Allah – their way. Sincere devotion to the truth and the belief in the brotherhood of prophets was not their style. Instead, they had a notion that Allah existed and was one of their chief deities. Therefore, they had many gods and the pattern of the true believers of old in submitting to the truth brought by the prophets was alien to them.

2. They did have a form of worship but it was like a form of appeasement. Offering tidbits of worship from time to time is like throwing a dog a bone to keep it occupied. If most of the meat has been eaten from the bone, then better for them and their various gods should not be stingy but should be happy for the notice and tribute paid to them as paltry as it was.

3. They had a perverse feeling about their culture and tribal loyalty that made them dull-witted but of course in their eyes they were gems of the human race.

In all honesty modern man seems to have advanced past that stage but the refinement of his beliefs still has the shadows of the past. The real challenge is to face reality and work at it. So it is imperative to seek out knowledge, the knowledge that does well and that brings a harmony of understanding without arrogance – a type of knowledge that can be tested without bias or cultural interference.

Probably the biggest uproar from this book is that concerning the Fifth Age of man and the gift given to him by the Creator Lord. What is that gift? To put it simply it is Islam. Some people might ask, "What kind of bloody nonsense is that"? "What, no gold, jewels, lofty mansions with heated swimming pools or pretty girls? What kind of a gift is that"? That is a gift of mighty import. At best, people can claim that this 'theory' is not a real fact but if they had really studied their own records – studied them as they should have been studied – this thought would not seem so strange. The fact still remains that a theory is just a theory and no amount of theorizing can replace the action with the deed.

What is this theory based on? The basis of the theory lies in several facts. The first fact is that there is only one God. The second fact is that there is only one brotherhood of prophets leading to the knowledge that in reality there has been only one religion – but the expressions of the same have been slightly different as to the peoples it came to, the language used and to a certain extent their rituals. The third fact is that the Quran and the Sunnah are inviolable up to the Day of Resurrection and that anything that contradicts these things is wrong. The fourth fact is that prophecy ended around one thousand four hundred years ago and that no more prophecy will enter into the world from that time until the Last Day of Resurrection. That is that prophecy has been completed with the passing away of that mighty soul – Prophet Muhammad (pbh) and no more prophecy and no more prophets will be given to the world. The fifth fact is that there will be a Resurrection of the dead for the Day of Accountability, known as the Last Day of Judgment, is when Paradise and Hell-Fire will be opened up for those deserving it. The sixth fact is that a heaven on Earth will exist before the great and terrible day of Resurrection but it is not everlasting and therefore only temporary. Since it is not eternal, it is not the heavenly Paradise. It is just a gift from the Creator Lord.

Many still cannot understand two simple things. They cannot understand why all of the prophets and all of their true followers were called Muslims and that they all followed the same Way and that the grades or ranks of the people are based on the truthfulness and sincerity of that one Way.

Now the problem with this theory even after it fulfills the essential six facts outlined above is that it can only push the envelope of credulity so far. That is, that it cannot come forth with new material but must remain hitched to already given material (open material accredited as valued literature in the perspective fields of acceptable religion henceforth known under the Abraham traditions of the Monotheistic class). In other words, alien records of dubious or perverted construction need not apply.

So the problem with any theory is the understanding based on the ability or failure to understand certain concepts especially if these concepts are unifying in nature. Why is this so? Some people want a special 'god' to deal with their special feelings and have been known to go out of their way to invent their god and his actions in order to buttress a good feeling about themselves. God loves a person to be kind. If one doesn't belong to 'my' religious belief system, then one's kindness is of no account. That is a very strange thought which borders on the arrogant.

It should be obvious that all true religions speak very clearly that there should be only One God. So why is there so much discord and hatred about it? Obviously, if one can't come to an agreement over the simple basics, then there must be something terribly wrong. There is and that is that the ego of man keeps getting in the way and he refuses to see anything that would diminish his so-called greatness. In other words, his perverse love of self is held in higher esteem than love for his Creator.

It is man's destiny to search for knowledge. It is man's destiny to be a vicegerent. It is man's destiny to learn lessons. It is many of men's destinies to try and escape accountability. What a prideful lot we have become.

Knowledge must be consistent across the board. After all, people have their 'records' given to them which are required reading by all. And the wisdom and the practices in the daily life have been required by all. No prophet ever came to tell people to do as they pleased or to only give 'two bits' to their Creator Lord. In fact, every prophet gave specific details on how man should work, work, work on those ideal building principles that will lead the soul to the right path and then to salvation. So the result is that one should practice it (the faith) because true faith is not a tongue wagging process but rather an external and an internal way of life.

One of the best modern books for understanding the Quran is Sayyed Abul A 'ala Maududi's Tafsir (pronounced Tafseer) called, *The Meaning of the Quran*. These volumes represent a refreshing and distinct look at Islam. However, like all books except the Quran, it has a major flaw and that is that Maududi combined two distinct occurrences into one occurrence at the same time. That is he put heaven on Earth on a flat, mountain less Earth whereas that flat mountain less Earth shall be after the kingdom of the Messiah is no more and not before this event. These things will, Allah Willing, be explained in further detail in the chapter *THE GIFT OF ALLAH TO MAN*. It is important that the Divine Records come to agreement and with the help of a coordinated effort of Quranic verses and some hadith material these things will be better understood.

It has also been hinted in Maududi's Tafsir that Islam is limited to certain areas. This may sound odd at first but it is completely true concerning the spread of Islam as to controlling certain populations under the Shari'ah Law. According to two of the four rightly guided caliphs, the control of strictly Islamic territories known as Dar al-Islam will be limited in scope. Now that doesn't sound like a religion for all peoples and for all times. However, the Six Ages of man theory explains that thought and puts it into a proper perspective especially when dealing with the Fifth Age of man. (See appendix D).

Strangely enough, the above thought is quite similar to

the idea of souls asleep in Christ or the Messiah-soul. People can ask, "How can these things be?" To be sure, these things are difficult to understand until one looks closely at some commentaries on the Quran by Abdullah Ibn 'Abbas (Allah have mercy on him) in conjunction with other hadiths. Simply put, the flat out overall victory for Islam will not occur in the Fourth Age but will occur in the Fifth Age and that is the age that the excellent souls will want to be born into except for those who by the Grace of their Lord are exempt.

All of this does not lead to a reminiscence of former scriptures from the People of the Book. For example, it might now be possible for some to grasp a better idea of why so much awe and adulation of the Messiah is put in the New Testament especially when the Six Ages of man theory is understood. It may also lead some to understand certain words and phrases in that book to be put into a proper perspective but because you (O Muslims) have The Book and the Sunnah you have The Future so it is NOT necessary to dabble in suppositions of the past peoples and what their records may or may not mean.

And these things are made clear because the time approaches swiftly of that which must come to pass so that the people can be gathered and the son of death will enter trying to lead one astray and the son of life will kill him and then will be established that blessed garden of paradise whereby souls can be purified and refined so that they will die and be given lightness on the Day of Accountability and will then enter what no mind has seen or can perceive of the joys of Eternal happiness in the real Paradise that shall endure for ever and ever.

According to the Six Ages of man theory, the people will be judged on the standard of Islam. It is Islam; specifically the Quran that will be the criterion used for judgment and this is validated from a correct hadith by Prophet Muhammad (pbh). It is he who has been preordained from before the foundation of the world to be the obliterator of unbelief and that is one of the cornerstones of this theory. It is not what is now found in the New Testament as to the words of Jesus (pbh) that the world does not know him for the world has heard of him. Say it is rather that the worldly does not understand and therefore is ignorant of him and that truth which he brought. As to him being preordained from before the foundations of the world, as was Jesus (pbh) according to the New Testament, what do you say? I say in remembrance of the hadith concerning the argument between Adam and Moses that Allah does not make snap decisions but has created the Perfect Plan though most men do not realize it.

HE IS NOT MY ANCESTOR

In today's world people are being overwhelmed by knowledge, news and materialism in all its varied forms. This has a tendency to overwhelm and dull the senses. What happens is that such an overload causes information to go in one ear and out the other and an attitude of 'so what' becomes prevalent. Just look at the nightly news and people can become overwhelmed and not think too clearly where we are headed. The news has given us instant hot spots, carnage and turmoil, which the brain tries to digest. Most of us can't digest it very well basically because the world is so dynamic that it is explosive.

Among the knowledge that bombards man is the knowledge of science and the multi-complexities of new thoughts and recent discoveries. Science in all of its splendor and glory is basically of two kinds: practical and theoretical. It is the theoretical branch of science I now turn to and specifically that branch of physical anthropology called paleoanthropology.

Behind the scenes of paleoanthropology are things like grants, money, the Darwinian Theory, Madison Avenue hype and a dainty tightrope walk around the explosive issue of religion. So what else is new? This is a part of the modern world and a part of life. The dedicated men and women who go out into the field are hardworking and most are concerned with exploring man's so-called distant past. Paleoanthropology has come a long way since its inception in the early 1890s from the digging up of some bones and guessing what they were and who or what they belonged to. Nowadays, the paleoanthropologist has a large array of tools that he can bring to his job. These tools are very helpful and instructive and include an interdisciplinary approach that helps the scientist to explore beyond just giving a bone a crude date.

The *National Geographic* show concerning one find in the country of Chad helped put Paleoanthropology on the map. This show was about hominid evolution and the search for man's ancestor. Basically, the hero of the show was a computer simulation of what Sahelanthropus tchadenis, nicknamed Toumai man, looked and acted like. Although this creature had chimpanzee characteristics, it also possessed enough characteristics of 'human-like' attributes that the creature seemed to steal the show even though there wasn't much of a creature to examine. The computer simulation was nice even though it gave off a cartoon-like effect. So the show was a big hit especially with a multiplication of uses of various scientific fields, Hollywood computer graphics with a plot to its story and a whole lot of conjecture smartly packaged to attract and delight an audience. What else can one expect nowadays?

It was all in all a good show and the conjecturing was fairly logical with lots of science thrown in for good measure. Toumai man is a creature that existed around seven million years ago. When a morphological study of its head structure was done, it was found that this creature most certainly walked upright and in order to survive in the area that he came from back in those times, it had to have some form of intelligence just to avoid the large predators.

Although we don't know much about Toumai man, the *National Geographic* special did do a good job at looking at various theoretical aspects of the so-called chain of human evolution. The most fascinating aspect of the show appeared to be the change of theories those paleoanthropologists were willing to share with the public. This had nothing to do with Toumai man but rather the sub-human evolutionary tree as they might call it.

Those scientists work hard off and in the field and sometimes they might let their hair down so to speak and give a more balanced view of what it all means. They don't know what exactly it means but their own theories, surprisingly enough, give a clue.

Admittedly, these scientists have, according to their words, only a small pickup truck worth of bones of prehistoric hominids. That is not much to go on in developing a sound theory but does, in the form of modern science, represent a significant breakthrough. The scientists would like to have a hundred more small pickup trucks filled with ancient bones supposedly representing the 'tree of life' of our so-called ancestors to fill in the gaping voids sufficient for their needs. But it is sufficient with the use of modern techniques of science such as it is. Thankfully they are still in the hunt for even more bones that could basically tell us what we should already know.

As fascinating as that may sound, it still is not getting us very far along the road to understanding the so-called ancient ancestors of mankind. What would help in this endeavor would be to understand mankind himself, for in truth that is where all the problems lie.

Let us start with a clear picture of what is really going on. In that way man might be able to expose a more complete understanding of his past. So let the scientists conjecture and let us conjecture and see what happens.

No matter how people in the various sciences crow about their accomplishments, it has been proven over and over again that an interdisciplinary approach to getting the bigger picture is the best way to go. Think of the process of a team rather than an individual. Now this team brings to the table various pieces of the puzzle so that a bigger pattern can be formed. It is like the collective consciousness at work. So let us think in terms of the future and see if it can lead to the past.

It has been said that the knowledge of science doubles every ten years. So everything being neutral, one can say that science can become a savior to mankind in dealing with aging, drudgery, exploration, various ills, wars, famine, toothaches, headaches and the old age question of which came first, the chicken or the egg. Bunk! Something always seems to get in the way of the starry-eyed prestidigitation makers.

First of all there is the science of genetics. Within thirty to forty years it will be possible to take any species of monkey and get them to walk upright like a man. It won't be easy but in theory it can be done. Scientists can now put legs, eyes and an extra head on a fruit fly's back with gene manipulation. Getting a monkey to walk upright with human-like hands and feet by gene manipulation is not so far fetched as one might think. Also, the attempt at increasing the brainpower of the monkey by a power of ten will also be in reach of the scientist. Of course selective breeding has to take place to ensure gene purity is maintained in this hypothetical bizarre creation.

Next, we already know of the chimpanzee training taking place by the aid of the computer. A chimpanzee's self-awareness is measured by the aid of interacting with symbols shown by a computer. Chimpanzees have been designated as having the capacity of a three-year-old child and can be trained to express their desires by using a computer with simple pictures.

Finally, the chimps need to have a modified voice box and be taught a language like English, at least in its rudimentary form, so that a very basic form of communication could take place between 'master' and servant. That would take some doing but it can be done. Put all of these exploits together with the understanding that science is closing in on how to boost the power of the human brain manifold so why not tinker with the monkey's brainpower?

Genetics has been given a big boost in the belief that it can transfer accumulated knowledge, eventually, in enhancing human capacity in the search for near immortality by holding out promises that some day the human life span will be quadrupled, the power of learning will be enhanced manifold, people will be made stronger and more beautiful and all sorts of diseases including the common cold will be defeated.

Either the *Discovery Channel* or the *National Geographic Channel* had a program about space where an oriental man, presumably a scientist, speaking perfect English so he is presumably an American, waxed high and mighty on what man can do about populating alien planets in the future. According to his scientific opinion, one day man might alter the genetic makeup of the fetus to produce a being that could live and work on the planet Mars. The computer version of what this being might look like was presented. Quite frankly, 'it' looked like a tree trunk with a peculiar head and long, thin arms and legs.

No doubt this super being would be able to withstand the climatic rigors of planet Mars and the solar radiation booming down on it. People should be allowed to see this program just for the shock value of seeing our futuristic thoughts being put to use. It does present quite a thought about what mankind thinks he will be capable of in the not too distant future. That thought may go over big in some circles but it is kind of scary to others. This idea is not a new one and without letting the cat out of the bag too soon, it kind of sounds like part of the human race is dizzy with thoughts of being likened to a god.

Man can dream can't he? Yes and man dreams of many things – things that promise him a step towards immortality and a chance to become a 'manipulative divine'. Warning bells should be going off all over the place on this idea but science has a truthful out. Man in general wants to improve his condition and without pushing the envelope of knowledge further and further back, how is man going to find the super cures for diseases and making man a better created being? After all, what is wrong with genetically enhancing man's brain so that he can utilize more and more of what he has? Scientists have found out that drugs are primitive as compared to possible genetic manipulation so genetic manipulation would seem to be the wave of the future.

Scientists, it can be argued, are basically honest and levelheaded. So aren't we all but history paints a slightly different portrait. Experimental humanization of the monkey, of course under highly controlled circumstances, might even aid the development of our own understanding. This sentiment has been reported in the news and therefore gives scientists an excuse to do certain kinds of research. So can it be argued with the invention of the atomic bomb and the benefits the atom can have for mankind. Look what headache that is now turning into.

Without a doubt, scientists with their super computers and advanced experimental research are pushing the fold of knowledge to super advanced levels with lots of hope and desire to be yet fulfilled. One can always use an intelligent monkey. Besides, an affordable trained drudge like an advanced monkey would free mankind from dull and boring chores and be more like a pet companion than a mere robot.

Don't think that scientists can't do it given the proper time and funding. They are already trying to unlock the secrets of our genetic code and are coming quite close to making major breakthroughs on genetic and cellular activity that can help, as can be argued, in the search for many cures for mankind.

Well, as has been said before, "There is nothing new underneath the Sun". This should give us our first clue as to what could have happened in the distant and forgotten past. The Quran shows us about repeating patterns as things are done again and again. Different peoples and conditions were involved but the patterns remain quite startlingly recurrent.

The Pape

The Pape is a completely imaginary creature that never existed and will never exist. The word Pape is an invented word that has no meaning whatsoever until one is given for it. The word Pape is composed of the letter 'P' which stands for people and the letters 'a'-'p'-'e' that stand for ape. Hence the word Pape comes into existence. Our imaginary Pape is going to help us find a solution to the so-called ancient primitive ancestor that paleoanthropologists seem to want to throw at us. Now even though the Pape never existed, the paleoanthropologists would wish that it had. Why? The reason for their wish is that it would solve all of their problems and get the monkey off their backs.

This imaginary creature looks acts and behaves like a monkey but its origins are not from the monkey, ape or any such animal. It coexisted with the primitive monkeys and apes over twenty or fifteen million years ago but it always walked upright. How it got on planet Earth is a mystery since it didn't seem to evolve from any known creature. That is beside the point. The point is that the Pape is our ancestor and not the monkey.

That concept is plain outright silly and it is silly for several reasons. First, that creature never existed. Second, it doesn't conform to evolutionary standards before it came into existence. In other words, it just popped down out of the sky in all of its primitive and hairy wonder. Yet this imaginary beast-man would answer a lot of questions and get the scientists out of a tight corner that they put themselves in.

What is that tight corner that they put themselves in? The tight corner that they put themselves in is the suggestion that man is totally of this Earth and therefore has evolved from creatures that were monkeys or apes. These monkeys or apes being totally inhabitants of this planet evolved from other creatures and so forth and so on such that before a certain creature climbed out of the 'primordial' slime, it evolved from that primordial slime. So, if our ancestors are from the primate category as in apes or monkeys, we are then the future evolved beings derived from pond scum. Why is that so? It is nonsensical to just stop at the monkey level because it makes more sense to go further back in time and become acquainted with our most ancient relative – pond scum.

These scientists should be responsible for their theories and take a look (a professional and responsible look) at what they are saying and what that implies.

They may in Western countries imply that men, and the prophets were only men, were descended from pond scum. They might even twist and twirl by declaring that Jesus is only half related to pond scum since he only had a mother but no father. Why cover the obvious logic that their theories would imply?

A close look at their theories of human evolution gives one pause to think. Every so often they have come up with a minor theory and then while it is running its course, they change their theory for another one (based on further knowledge), which they hopefully can be satisfied with. Then onwards they leap into another theory hopefully more advanced than the previous one.

Their latest theory is quite amazing and is getting closer to the truth of the affair. Before their latest major theory on the development of the ancestors of man, they pictured the evolution of man like a race whereby one species of ancient man (or animal-man) hands off a baton to the next species of ape-man and so on and so forth until modern man was born. That theory held for a while but now it has been sent to the graveyard. Their latest theory is based on a peculiar finding as reported in the program done on Toumai man. This theory is based on at least two or three types of walking ape that were found to be living in approximately the same area and at the same time as each other. These records don't lie so indeed there were several species of ape-things living together at the same time. This would tend to mean that at various places and at various times different species of apes, monkeys or whatever came down from the trees and started practicing strange and abnormal behavior. This idea was put forth in my first book, *The Shining Light of Islam*.

So what does this new theory actually mean? First of all, according to their own admissions of the dearth of material found, they may actually find more than just a few species of apes/monkeys doing this odd behavior together. Hence, it is possible to have five or six species doing this and not only just in that little area they investigated. Why? The answer is that finding these 'representatives' of supposed human evolution in bone form is exceedingly rare and that they have after many decades of intensive search found only a few samples of bone.

What does their latest theory translate in terms of common logic? Well, for one thing it could allude to something like the Hollywood movie Killdozer whereby a meteorite that had a living essence fell to Earth and when it came in contact with the blade of a bulldozer, it caused the bulldozer to inherit that essence and then the bulldozer came alive and went on a killing rampage. If they don't buy that conjecture, then what about the next Hollywood flick? Hollywood refined that theme in Stanly Kubrick's 2001: A Space Odyssey. In this show an outer space intelligence in the form of some block sent off signals at a critical time when two forces of ape-like creatures were about to do battle. One of the groups got the vibes from this mysterious block and defeated the other group of ape-like creatures by having a form of intelligence printed in them. If they don't buy that explanation, then they are running out of room to maneuver.

As explained somewhat haphazardly in my first book, it is just impossible for these creatures to successfully get out of the trees and start learning to use their new found magical walking abilities and survive the carnivores. BUT THAT IS EXACTLY WHAT HAPPENED!

Now monkeys and apes are pretty clever and there is no denying that. So at various times and various places, they seemingly decided to start on the road to becoming human. But they don't want their advanced cousins (humans) to know what they were up to. That is why they only came down from the trees and started their experimentation in the dawn of prehistory. They wanted to keep their change of character a secret from the human race. That tomfoolery explains why the apes and monkeys don't do anything so foolish today. They don't want humans to see that they will have any future competition. Or perhaps these animals don't want to be that stupid again and degenerate into human form.

If people are a bit confused, perhaps we can look at the picture from another angle. If we are related to pond scum, I suppose a flower is a more advanced form of life than that slime. Will this make simple gardeners afraid to go into their gardens? Could it be that the plants are plotting to wage war against our superior race? Will there be a re-make of the Hollywood movie Day of the Triffids whereby some sunflower – looking plants will start moving about and attack man?

Scientists and their theories on evolution (evolution does exist) have made some wild guesses that are down right comical. Since they have pegged an ape or a monkey as our relative, they must answer a certain question. If we are men/women and sprang from an animal, does that make us an animal? If the answer is no, then what does that mean? It means that by some form of magic an animal has somehow become a human or a different form of life. If the answer is yes, then what a wonderful thing our little minds have made – we are animals. Let's see where that leads.

Instead of calling Jesus only half pond scum (one should go to one's roots of creation if one is not embarrassed or humiliated), perhaps these wizards of science can get away with calling him an animal. One mad scientist might proclaim, "I am waiting for the glorious return of that animal." This is nothing but idiotic lunacy! We are not animals although, come to think about it, some people do behave in a peculiar fashion and have certain characteristics that would embarrass the animal kingdom.

When did the animal turn into a man? Since there is such a thing as reverse evolution like the whale, when does the man turn into an animal? Scientists have been setting certain criteria for this question ever since they have believed that we crawled out of the swamp. According to one human or perhaps one animal, former president Harry Truman, said, "If you can't stand the heat, get out of the kitchen."

Various theories have been put forward as to what makes an animal human. According to one genetic researcher as shown on the *National Geographic Channel*, being human is not really walking or talking like a human. It is only when the 'creature' has the ability to express him or herself in artistic form then he can be considered a human. Perhaps he knows something that we don't know. Perhaps Jesus was a painter or an interior decorator?

We are not animals! Scientists proclaim that it is not their job to go into a philosophical debate about this issue. Are they trying to tell us that thinking is not their job? Their job supposedly is to make theories based on a limited form of cognition. Truly, their cognition is limited. It is limited especially when they don't use all of the tools they could use and pooh-pooh religion as a source of investigative science.

All of this was said to bring about a contrast. These people dabble in religious fields by making pronouncements about our creation but do not feel that they should shoulder any responsibility for it. It is as if they can make pronouncements wholesale about an area they are ignorant about and then come away smelling of wine and roses.

The danger is that these scientists and they are admittedly dedicated and hardworking, have by implication thrown out the concepts of the One God, the one brotherhood of prophets and the Last Day of Judgment. Of course they didn't do it directly but the senses are so dulled by their speech that one can conclude that idea by taking in the trend as to where our supposed super species is going. Now when the senses become so dulled with this so-called scientific, highly advanced way of thinking, one can lean towards the feeling of putting one's trust in the courageous and wise men and women who have been serving up such tripe to the world.

Religion is a science also and some of those religious people have been accused of, and rightly so, in serving up their own foolish tripe and in such a degrading way as to make the mere mention of religion a mere sport fit only for confounded and confused individuals. So in a bizarre way the scientists dealing with the subject of man's evolutionary chain have some excuse, feeble though it may be, in manufacturing a system devoid of wisdom. It is like tit for tat.

It would be nice to try and fathom an understanding of man's role in the universe to some extent. Even though we can't reach the totality of answers, we can find enough evidence to land us on solid ground. It is absolutely true that God created man and that he did not evolve from any animal including the ape or monkey. No book or piece of literature is going to give us a step-by-step description of such detailed exactness as to entertain the total picture, however. Yet the evidence is there nonetheless.

The best way of approach to the of study of man and his ancient past is to look at <u>modern man</u> and see if he has tendencies that form certain patterns that can be studied. If man was once a flower or pond scum or if he was once a long armed hairy animal that swung through the trees, then a look at modern man would give little evidence about his past. However, if man is from the beginning a man and only of that type of creature, then his tendencies, no matter how tenuous, can be studied and from this one can learn a possible explanation of how the apes and monkeys actually became motivated to come down from the trees and doggedly practiced walking on two legs while developing from there. It might also show how experimentation in this mighty foolish feat continued at various times and in various places.

The sciences of psychology, history and religion need to have their input in the concept of the genesis of man along with other scientific fields. Who could be so arrogant as to close down the avenue of understanding by claiming that only their voice should be heard? Just because a good paleoanthropologist is a poor psychologist doesn't mean that he can dismiss the science of psychology. A wise man will utilize any avenue of approach that can bring useful suggestions to his specialty that in the end result will allow him or her to draw a better and more balanced understanding. The basic, simplistic ideas are there and in enough quantity such that if one is looking for the basic truths, then one can access them.

This is admittedly a very difficult subject. Perhaps a list of understandable factors expressed in easily digestible parts is the best way to approach this problem. The reason for this is the very complexity of the issue at hand. With this being the case, it is better to break down the subject into manageable parts that can reflect the whole. At first the list to be given will seem to be quite on the odd side but it should be remembered that the human soul and its psyche are very elusive subjects of and by themselves. This is what a partial list of what we need to know may look like.

1. Man's Essence
2. Awareness
3. The British Empire
4. Fallen Angels

The above list looks like a jumble but it does represent a beginning of understanding. The best way of looking at a subject is through a thorough study of the Quran and hadith literature. This represents the foundation of understanding. After that one can use real life examples or thought processes and analogies to center in on the topic at hand. Because the topic is a conjectural one but still concrete because of the Word of Allah, the ideas can be expressed in areas common to the general public so as to open up a kind of corridor to understanding.

The truth that the Quran is pure and is the unadulterated Word of God is undisputed among men of intelligence. When it is viewed under the most stringent of tests, it does not weaken in its claim to be 100% accurate. This Book has had many detractors over the centuries who have presented pathetic excuses for believing otherwise but their lists of debatable attacks have all fallen by the wayside and have been exposed as unscientific, petty rivalries, misconceptions and plain outright lies. This Book's consistency when looked at in its Arabic, lexical form shows no weakness at all. Only self-vested interests and cultural denigration has caused the Quran to be held up for ridicule by perverted human beings.

All of the creation has what is called an essence. Rocks, trees, insects, animals and human beings have their own individual essence. That is why stones and trees and so on and so forth can have a form of speech. In other words, they can actually talk when commanded to do so.

It is important that a case for anything that is said about things that are seemingly strange is that they must be built from a perspective of understanding not so much by blind faith. Therefore, it would be advantageous to help our case for understanding if certain 'facts' can be supplied from either legitimate records or legitimate experiences and then to incorporate these things into a comprehensive whole. This is what a true scientist does when he finds a bone and then goes to various disciplines seeking help for a comprehensive examination on what he has found. So the wise scientist will go outside his field of study by applying all methods at his disposal to get a clearer picture of a more complete and actual event of the story he is trying to build. This is what is being attempted here.

Now let us take the case of Solomon and the ant:

At length, when they came to a valley of ants, one of the ants said: "O you ants, get into your habitations, lest Solomon and his hosts crush you (under foot) without knowing it."
So he smiled, amused at her speech… Quran (27: 18-19)

Now the ant's brain is extremely small and its physical

communication is very limited as well as its eyesight. It does not communicate verbally for it has no vocal cords for speech. Its life force (essence), however, can 'talk' but its life force has nothing directly to do with the brain. Take away this life force or essence and the creature will be physically dead. It is this spiritual essence or life force that causes a creature to be responsive to the environment as it was planned to be responsive. (See appendix B).

Solomon's words to another life essence (the Hoopoe bird) are not one of pleading or begging. They are words of one fully in command as a vicegerent to a servant – dominating and powerful as well as able to control and punish. In this case, the punishment is not a slap on the wrist but the highest form of penalty – control over life and death. So at least we are shown that a vicegerent is not a wimpy ruler or commander of forces. It is true that Solomon was a prophet and had special gifts but he was also very much a human representative of what a human being could be like. Of course his ability to be a true believer and to be true in his thankfulness to the One God goes without question. All this shows is that man had the ability, at one time at least, to have the potential to show his quality of rulership as being a custodian over a host of created things.

And he took a muster of the birds; and he said: "Why do I not see the Hoopoe? Or is he among the absentees?

"I will certainly punish him with a severe punishment, or execute him, unless he brings me a clear reason (for his absence)". Quran (27: 20-21)

When the short lived speaking in tongues were permitted to exist among some of the followers of Jesus (pbh), this occurred not in the brain but in the link between the various essences that were involved. When Paul, for example, was bitten by a poisonous snake, the poisonous essence of the snake's venom refused to activate. In other words, toxic chemicals were introduced into his bloodstream but they refused to do what those toxic chemicals do and therefore Paul did not die or become sick.

This essence can't be caged and studied like a laboratory rat so it is only in front of us as a vague notion. Lots of questions can be asked about this essence but it is best to leave what can't be known alone. Some hadiths deal with this issue but in a basic way but these hadiths generally refer to the balancing out of justice. It should be said, however, that the essence in all things are under what is called Divine Destiny and should not be measured as a 'holy' thing or some type of magic.

The Solomon (pbh) and the ant example does show several things. It shows that a human and a non-human life force have linked up and not in a normal, everyday way. This is where the idea of awareness comes into play. It is not just the awareness of one created being (Solomon) but the awareness of the ant and its fellow ants. Solomon could not hope to hear the squeak of a mouse under the din of his marching forces much less than an inaudible murmur from an ant that doesn't have a voice box or a brain to produce such speech. It can only be guessed that this share in the essence of life or even non-living material such as a stone has some form of a pre-planned existence from its Creator Lord that generates types of invisible waves.

These waves are independent of the body physical of the life or non-life form and are arranged in a pre-planned creation of some form of existence from whence they derive their being. That would mean that there are categories of life and non-life forms coming into existence and that makes sense when one understands that 'they will bow down willingly or unwillingly' as it is stated in the Quran. Or another example from the Quran is the form of different rocks (inanimate objects) but still they have a certain awareness of giving off praise to their Creator Lord.

In other words, gold knows itself and doesn't of and by itself suddenly become iron or parade around as water. Hence, the creation knows its Creator and conforms to a pattern of essence-creation and the form it has been eventually allotted to it. This in Arabic is known as a form of wahi (pronounced as wahee).

The supposedly changing of water into wine by Jesus (pbh) would be an alteration of the essence of water to another essence done by command from a superior creation (man over matter - vicegerent) to that which is an inferior creation (servant). The very important lesson here is that this was not magic but an understanding of the forces of what some have called nature or the so-called secrets or forces that binds the universe together in a form of inner connected oneness translated in Arabic as a form of tauheed. This then reflects the order and Oneness of the Creator Lord as a Sign of His Unity and Uniqueness.

Anyway, the whole discussion revolves around a deeper understanding of the word vicegerent. Mankind was made like their father Adam (pbh) and as a vicegerent he was put over the Earth as a dominant controller. That means that he is the highest of forms not only essence-wise but physically-wise on Earth. Therefore, a true vicegerent commands while other life forms must comply. While the Angels bowed down to Adam, thereby showing his elevated and superior status as a commanding being in the heavens, on Earth this form of bowing down would show that the Angels would and must show compliance to man's will.

All of the above makes sense because a true vicegerent cannot rule if he is abused, ignored or treated like a buffoon. The following is a direct quote taken from The Meaning of the Quran (Tafseer) by Sayyid Abul A'la Maududi p.68 note #45 concerning verse # 34 in Sura II:

This was symbolic of the submission and subjugation to Man of the angels who manage the Earth and that part of the universe which is connected with it in any way. As Man was being appointed vicegerent by Allah's command, it was ordained that all the angels, who worked in that part of the universe, must, as far and as long as Allah willed, co-operate with him in their own spheres whether he wanted to use or abuse the powers given to him. This implied: "You must help him in whatever he wants to do, irrespective of whether it is right or wrong. Supposing he wants to offer prayer or do any other good work, you must contribute to it in your own sphere. Or, if he wants to commit theft or any other evil, you must be helpful to him, as long as We let him exercise his powers in that way. But when We revoke those powers, you should cease to co-operate with him." This may be illustrated by the example of a government officer. He is obeyed within his jurisdiction by every official but no sooner is he deposed by the government than those very officials who carried out his order, cease to obey him. They even handcuff him and take him to prison, if the government so orders. It appears that the angels were ordered to have the same kind of relations with Man. Possibly the word Sajdah, "bowing down", here may be merely symbolic of submission, and probably the angels were ordered to perform some such act actually to signify their subordination.

By looking at the world today we can see knowledge and leadership in action from the human beings scattered across this planet as the term vicegerent takes on peculiar forms. The dictators, the drug lords and a whole host of other bad guys are mixed up and set into the fray with good people just trying to do their jobs as normal, regular people of any race, religion or political persuasion. This must be so with a world that has become dynamically diverse in life style and purpose. **The people only know the outward aspect of the worldly life,**

and of the Hereafter they are heedless. Quran (30: 7)

We as humans can't all at one and the same time be on the same path because our very intentions, desires and purposes are so varied. Basically, with our different viewpoints and perspectives, we are a ship with a badly damaged rudder sailing through an immense ocean. The Creator Lord gives us these experiences in that we should learn about Him and His system not because of our wealth of goodness or because we have earned His approval (He is too Mighty for that) but because, in part, He is the All-Merciful, the All-Kind.

We can respect our bodies and our minds but can we respect our true essence or life-force? That life-force does not decay nor does it turn into dust. It is an essence that must acknowledge His Essence as Lord and Creator but our essence is not the same as His Essence so we must not assume that a billion of the best of souls could ever even begin to compare with Him.

It has been said that the first man, called Adam, was molded from the Earth in the form of clay-like mud. Then, after being a lifeless form, some of the Spirit was breathed into him and he became alive. Whatever some of that Spirit is we don't know but we can know that it is not a part of the Creator Lord or any other such fantasy. What that essence did was to make man a very powerful entity among the creation – an entity that unlike the Angels would have a free but limited reign to mix with the 'lower' elements of creation. This then becomes the key element in the study of man.

Man now becomes a potential ruler and controller with many of the Angels being held as servants and helpers of a specially gifted creation – a specially gifted creation that had an innate form of wisdom but not a creation that could claim divinity. Hence, the potential existed for corruption if wisdom was not applied and thence the fall of this creation into disrepute. Knowledge could be applied but knowledge without wisdom is a very dangerous thing.

It is to Adam that we now turn. He was a complete and perfect creation in form and essence lacking nothing and being made complete in all things including his soul. There was no sense of want in Adam in that he missed a counterpart because his counterpart was in him in his completeness. It would hardly be fitting to say he was man/woman but rather say that he was complete in himself. However, the One Who creates is the One Who plans for He is the Evolver (Al-Bari) and He knows that what is before His creations and that which is after His creations and what will become of His creations.

Mankind came into being when we were taken from Adam's back and like our father we had that vicegerent essence instilled in us.

It is He, Who has made you the vicegerents on the earth and raised some of you above others in ranks so that He may test you in what He has given you. Indeed, your Lord is swift in inflicting punishment yet He is also very Forgiving and Merciful. Quran (6: 165)

Things, however, are formed in stages just like Adam. He was not fashioned into completeness at one and the same time – body and soul. So the souls of men, once they were called out from their father Adam, would conform to the idea of stages. What that means in full of course is unknown but simply put into an example that can be known is that men do not come into the world at one and the same time. They come (are born) in waves or groups thus completing the Plan of the Creator. Even so, all souls that enter into this life were imprinted with the instruction for obedience from the beginning but having a limited free will to choose the pathways of action.

A tradition of Ubayy-bin Ka'ab – All of mankind was assembled in <u>separate</u> groups giving them the understanding that they were to be made vicegerents on Earth while imparting their wisdom and understanding, i.e. granting them power such that each individual (would be) completely and fully responsible for their actions. Therefore, no individual can absolve him or herself from the responsibility of deviation from Allah's Way. It became man's duty to make use of all of his facilities of the gifts given to them to <u>discern</u> truth from error.

We have indeed created man in the best of molds. Quran (95: 4)

Mankind according to this verse did not start off as animal nor did he originate from pond scum. In truth, man originated as a perfect creation able with perfect Fitra or naturalness of disposition to use his talents wisely and with an ability to think and understand guidance. However, being higher than the Angels, he had that free spirit or free will to investigate things with wisdom or to forgo wisdom and get excited about other things. So, his choice would influence his character as a spirit of creation and when and if he woke up to the Mighty Graces of His Lord, the errant soul could always turn to Him for Guidance, Mercy and Forgiveness.

So many empty spaces and gaps in our knowledge remain after this point but what can be surmised is that the soul of man is not physical but spiritual. A spiritual creature can sense the physical but cannot engage in physical acts. In other words, the spiritual world of men cannot smell the flowers, taste the honey or pick up objects of interest. In a certain kind of sense, a spirit can form awareness of its self and its surroundings but it can't enjoy or engage in materialistic affairs. That provides us with another important key for our understanding.

The spiritual world being higher and much more stable than the physical-materialistic world has the ability of enduring instead of decaying. Now the physical world with its earthly shells incasing this spirit in what we recognize today as the body can show us certain tendencies. That provides us with yet another important key for our understanding.

One of these tendencies was called the British Empire. The British Empire is just one of the many examples that can be used to pick out a system of actions between men. If we remember that men do not operate in an empty shell, then we can see that people in this saga were operating in a body with a mind and a soul and a will.

Basically, the British Empire was an empire that operated under a certain system – the so-called superior forces of worldwide smarts versus the underdeveloped systems of certain peoples that could be utilized for the enhancement of the <u>few</u> chosen ones. It wasn't the British population that grew fat and rich as much as certain individuals of that population. What happened is that a dominating force looked upon a simplistic force and this dominating force tried to assert control over any society it could get its hands on.

This was done through various means – some honest and some dishonest. The result was that goods and materials were produced dirt cheap on one end and were sold on the other end for a huge profit. It is something like, 'we will give you one dollar a month but we will make from your labors one-hundred dollars a day'.

Now this system laid railways, made roads and bridges and developed schools to show that their system should be respected and that it was a higher system than the indigenous population had. It could also benefit the local population if the British left for some reason but if not that would be those other people's lookout. The primary reason for introducing cultural advancement was not for the gain of the local population but for the enduring upkeep of the British Empire.

In the above example, one essence, as in a nation, overwhelms and dominates other nations through a process of superiority of knowledge, strength and belief systems which turned out to be fueled a lot more by greed than any altruistic philosophy. In other words, the spirit of domination was the driving force of overtaking and conquering what appeared in worldly terms as other systems that were primitive as compared to the British Empire. Some have called these actions as the law of the jungle attitude.

Truly man was created very impatient. Quran (70: 19)

The lesson here is that people can have the tendency to overwhelm and manipulate others to force them to alter their behavior and existent life styles. In fact it is much like a toddler who sees something and wants it to the embarrassment of his parents if they are in a store surrounded by people. That toddler has a mind and a soul. It isn't his shell of a body that causes the uproar.

The lesson here might not be so obvious with all the forces acting on the toddler. First of all, this is not the spiritual world. This is planet Earth. Secondly, the toddler must be trained as to what is considered appropriate behavior with all its limitations and responsibilities in this realm of worldly delights so as not to create a scene next time when he goes shopping with his parents. Thirdly, he must rationalize about the appropriateness of his behavior and either accept guidance or find a way which they often do to subvert parental guidance into getting what they desire. That provides us with another important key for our understanding.

This worldly life is but a sport and pastime. In fact, the abode of the Hereafter is better for those who desire to be safe from harm. Will you not, then, use your common sense? Quran (6: 32)

Our next step is the discussion of Angels. Like in all realms known to exist from the stars and planets to human societies, there are gradations or levels of things. One celled simplistic life forms and elephants have things in common but they are on different levels by classification. The same thing can be said of the Angelic realm. Most of the Angels we can assume are of a non-thinking or reasoning type of being.

They are given a limited purpose and they fulfill that purpose without any sign of weakness or defect. In other words, they are programmed for a certain job and might be considered unknowing when it concerns free rational thinking outside of their given specialty. That is to say that they are perfect beings created for specific jobs and they do those jobs to perfection without the least care for any intervening side issues. There are some Angelic beings who have been given immense thought processes and do think but not in what we would consider rational thinking. In other words, they are high class Angels, sometimes called in English as the Archangels, that are so perfect and dependable but with such complex jobs that they do have a special place in creation. Even with this distinction, they would never think of disobeying their Creator Lord but are rather so attuned to His Will that they might be called super enforcers of that Will.

They (the Angels) fear their Lord, High above them, and they do all that they are commanded. Quran (16: 50)

The conjecture about Angelic beings is just that. What we can know is that Angels cannot fall in disgrace. In other words, there isn't anything like a 'fallen Angel' in Islam at least. This goes against certain revealed scripture or even some questionable scripture but there may be a reason for this. If a person is writing about a difficult concept much like the one that is being written now, and is under space and time restrictions since he can only write a few verses instead of a book, then perhaps a code is necessary. Therefore, <u>Angelic-like beings</u> can fall whereas in truth there is no such thing as fallen Angels.

It is only conjecture at this point but when Angelic-like beings are placed into an important category as guide, teacher and watcher, the term Angelic-like beings may start to make a lot more sense as to what they actually were. The teacher, as far as we can know in this world, has a very respectable job.

41

A teacher is supposed to be a very experienced human being with a job of imparting successful insights to his/her students. The students of course are human beings like the teacher but come as inexperienced learners. We cannot consider these teachers as anything more than humans. So the idea of Angelic-like beings falling is quite different than the non-real concept of fallen Angels. One can surmise without attacking scripture that these fallen spirits are no more than our spirits but of course in the sense that a third grade student is no less a human being than a third grade teacher but with a whole lot less experience, awareness and practical wisdom.

It has been known that a few teachers hired for their abilities and management skills and professional maturity have been arrested for the unbecoming conduct of acting in certain depraved ways. Not only did they become unfit for the job, their behavior spoke volumes on how sick they were such that they ended up in jail due to their unprofessional and perverted conduct. This shocking evidence has been well documented and can be appropriately tagged as 'a fall from grace'.

Relate to them the story of the man who We sent Our Signs, but he passed them by: so Satan followed him up, and he went astray.

If it had been Our Will, We should have elevated him with Our Signs: but he inclined to the earth, and he followed his own vain desires. Quran (7: 175-176)

I have used these verses before in showing what a certain powerful individual could do and would become. Here, I am using these two verses in showing that there are categories of men who still retain a power but that power is used in a perverted way to entrap others and bring them low.

Let us take an example. Men belong to not only groups but also categories and they form a brotherhood as in a common bond and purpose whether it is the brotherhood of prophets or an established union for the protection of workers. If a person studies a man like Solomon from an in-depth look with the help of a Quranic Tafseer, one could get the impression that he is more than an ordinary man because of his great stature and powers. However, that impression would be dangerous because Solomon was only a man and no higher than a man. Certainly some of his deeds seem to be other than human but then only to those people who keep forgetting what a vicegerent, in its full splendor and glory, means. In fact, all of those prophets that have entered the physical world as well as ourselves have been veiled as to the greatness given to us.

That is a great thing or such tremendous havoc would reign on the Earth such that the Earth itself would be torn apart. It is not expected that man covered in his earthly form can grasp or even understand his creation in total nor can he comprehend the Wisdom and Controlling Authority of his Creator. He is not asked to do the impossible but he is asked and will be accountable for submitting himself before His Throne and no person will be able to wiggle out of or deceive the Creator Lord.

Let not those who deny the Truth delude themselves that they have won the game; indeed they are incapable of frustrating Us. Quran (8: 59)

Every being in the heavens and the earth will come to Allah as a servant.

He does take an account of all of them and has numbered them (all) exactly.

And every one of them will come to Him singly on the Day of Judgment. Quran (19: 93-95)

Common sense tells us to ponder over things to bring to our consciousness the amazing creation and the Divine Will in as much of the splendor and majestic glory our little insignificant essences can comprehend.

NO! But they fear not the Hereafter.

**By no means! For verily, this (Quran) is an admonition.
So let him who wills, take heed. Quran (74: 53-55)
Say: "O my servants who have transgressed against their
souls! Do not despair of the Mercy of Allah: for Allah forgives
all sins: for He is the All-Forgiving, the All-Merciful. Quran
(39: 53)**

We know in Islamic science that Angels cannot fall. It can only be speculated at this time and perhaps will be proven later that the category of 'fallen Angels' were actually men whose power, dominance and wisdom became corrupted in the heavens and then came down to Earth and corrupted humanity.

There is a law, for those who have understanding that goes something like this. One is accursed if one tries to enter the flesh without the pattern that has been laid down. Truly, this sounds like science fiction if one breezes over this thought without consideration. The basic idea found in Islam is that a man, in order to be a true man and placed in the category of the sons of Adam for salvation, must be born of the flesh from a woman. This includes the normal birth or even virgin birth. Otherwise, the soul that does things differently has willfully disobeyed Allah's Will and is accursed. It is accursed because it has rejected the Way and has tried perversion and can only be forgiven if it truly repents from its perversion.

Now people would find it fantastic that souls could enter the Earth plane without being born of woman or that they could by some neat trick form a body to envelope their soul. The truth of the matter becomes apparent when we admit that we really know nothing of the First or Second Age talked about in my previous book, and hardly anything at all about the Third Age of man. It is not unconceivable that a powerful but very corrupt vicegerent in heavenly form would pull such a stunt. Indeed! By following this line of reasoning, some of the mysteries that have long puzzled man and which have reputed to come from ancient books can now become clearer.

Now admittedly that notion sounds too utterly incredible and too fantastic to be real. Why is that? Because mankind in general did not take the time to study his God-given records as they should have been studied, they are left somewhat in the dark. Some will say, "There is no proof of this". But there is. One of the great prophets came and declared it. He not only declared it but he spoke openly and poignantly of the prophet to the whole world who would come after him. The prophet who spoke of these things was none other than Jesus (pbh). (See appendix C).

"Jesus answered Verily, verily; I say to you, except a man is born of water and of the Spirit, he cannot enter into the kingdom of God.

That which is born of the flesh is flesh; and that which is born of the Spirit is spirit." John (3: 5-6)

In other words, man will bow down to the pattern set by the Creator Lord willingly or unwillingly but he will bow down. Now the spirit must follow that pattern set from Adam or he will be accursed and it will not be accepted from him to be born in this age (enter into the world without a body no matter how powerful he might have become). He must have his soul (spirit) born of a woman – the watery birth in the womb, that is, of flesh or else the salvation (kingdom of heaven) will be forbidden to him. So the Angles take the soul of man and place it where Allah Wills it to be placed in the specific womb He Wills it to be placed after 120 days of conception. Therefore, what could have been done in the very distant past by some is rejected.

Some say with their mouths that unless one is baptized by water as a Christian one will not enter into Paradise or salvation. Hence, Muslims, Jews and all the rest of mankind are doomed to Hell. Are these people trying to tell the All-Knowing Allah what He does not know? Have they been

45

appointed as regulators of His Kingdom? God forbid!

The statement by Jesus (pbh) makes it universally clear for those who understand that it is not the sacrament of baptism that Jesus is talking about but the willfulness of some who have tried to bypass the system or pattern set forth by the One God. What some people can't understand is that the sayings of Jesus found in the New Testament are not always placed in a proper context but have at times been haphazardly inserted into a place just to fit them into their books.

It is true, however, that the sacrament of baptism did exist as a ritual for Jews and the followers of Jesus. This sacrament was only a ritual symbolizing the new birth as to the submission to Allah. It is much like the Muslims' holy black stone placed in one of the corners of the Kabah. It is a symbolic thing and not something of worship or a 'god'. It is also important to understand that baptism is only symbolic and not a step for salvation and thus should not be used as a condemnation of others. In understanding that, the oneness of the records as well as the oneness of the brotherhood of prophets maintains its integrity.

Another interesting point in this discussion is the overturning of the general theory of transmigration of souls. Although I personally do not want to touch this subject with the proverbial ten foot pole, it must be said that people should purify their thought processes with Divine Revelation rather than inspire themselves with questionable voodoo philosophies. The Quran is explicit in saying, 'that was a people that have past on before you' and 'you are not responsible for what they did' and 'no soul can bear the burden of another soul'. This makes it clear that mankind is not put through a constant and continuous soul transmigration process.

They were a people who have passed away (before you) and they shall be repaid for what they earned and you for what you earn; you will not be questioned as to what they did. Quran (2:

141)

Every human being born into the world took the pledge of honor from before its earthly birth and along with the soul it is given its Fitra. So what happened before there was the Law and before the prophets but not before there was mischief in the Earth should not concern anyone. Therefore, it is much more practical to deal with concrete realities. However, what Allah has revealed should not be hidden.

When your Lord drew forth from the Children of Adam – from their loins – their descendents and made them testify concerning themselves (saying):
"Am I not your Lord (Who cherishes and sustains you)?" They said: "Yes, we do testify (to that)!" (This), lest you should say on the Day of Judgment, "Of this we were never mindful."
Or lest you should say, "Our fathers before us took false gods, but we are (their) descendants after them. Will You then destroy us because of the deeds of men who followed falsehood?" Quran (7: 172-173)

So, why has there been a discussion about fallen Angels? Simply put, it may clear up an issue that divides people against each other and also to debunk a myth that could cause divisions. It also clears up the cluttered road in debunking the idea that has spread around the technological world that our human ancient ancestors were of the ape and monkey family. So in effect it is like a war. One side has come close to putting religion into an invented system created by highly advanced apes that are now called human and this dulls the sense of urgency and reality of the Day of Accountability. The other side which is looked at as scientifically primitive or slightly retarded compared with the people who are really 'with it', seemingly stumbles around without much to say or at least giving evidence that is seemingly held as nothing more than holy fluff.

Now the tables are turned somewhat and people have

some ammunition to throw into the fray and do not have to feel as if their cherished concepts need to be mixed with corrupt thinking thereby creating such a horror of faith that it becomes easy to discard it at the drop of the hat.

It is also a sign for those interested in salvation that the idea of salvation is a deadly serious topic and not a flimsy, philosophical one. It also shows that no one set of people invented the concept of the One God, brotherhood of prophets and the one Way but that this system has been around for a long time. It also shows that some people have tried to pervert this system by tampering with their records but that the system was rejuvenated and kept intact through the one revelation that man couldn't pollute – the Quran!

We can't know exactly what happened in the past but we are left with only two basic choices. Either the scientists are correct and by implication we were evolved from pond scum and then an animal and then the animal became man (how utterly pompous) who evolved a system of gods and then invented a one god to rule all or to put it quite simply, that man could choose out of foolishness to become an animal for selfish designs including making sport with the creation and then found himself trapped in his own perversion.

The second argument has grave consequences that most would rather not think about but the scientific community should not be left off the hook to prattle on and on and trap people into a new religion of self-desire, empty hopes and false dreams that contradict Truth and lead others astray.

A very close study of Islam shuts down these perverted thoughts and grants balance to all things. After all, it was promised through the other scriptures that the spirit of truth would come and render to men the truth. Those who understand the Quran know that he did not come to mankind saying, "Take it or leave it". On the contrary, the Arab community was built up step by step in a piecemeal way with constant reinforced examples in a step-by-step process. This in itself is a great miracle that no human can fathom. How can this Book be for all places and all times and for all mankind at one and the same time and at the same time function as a step-by-step development for the faithful? Any book doing that must indeed weigh a ton and surely would humble any mountain it landed on. Yet this Book can easily be made to fit in the hand of a small child!

Since human cruelty and the desire to dominate others did not originate in the 20th Century, mankind should be aware that the office of vicegerent is not to be taken lightly and is very important and comes with a high price of responsibility as well as a very great warning. There is no need to give a warning to the Angels as they are strictly obedient to their Divine Lord and have no free will. However, a warning was given to mankind because he could choose to go his own way and try to excel in evil as the person known as the Anti-Christ did.

On one hand there is the Light. On the other hand there is the Earth and the Earth is filled with fantastic wonders, delights and deceptions. What fun it must be to pursue these hidden delights and deceptions and explore these awesome events. Unfortunately it can't be done because the soul in spirit form can't take on the lower world without having the lower senses of the physical to enjoy the smell of flowers, the taste of honey and the joys of sexual reproduction. Besides, all creatures have their own life essences, which compared with man's essence, are very slight.

It would take an <u>evil genius</u> to figure out how mere spirit could impose on these life forms and sample the goodies that could not be sampled by mere spirit forms. Ah! What evil genius could give some spirits the key to show how a wandering spirit could sample earthly delights while expressing power and control and even have nature stand on its head if it wanted to? Why with this key a spirit could become a god and could command and manipulate with impunity. In fact, the partaking spirits could gather together and become like God Himself and vie for His Throne.

What evil genius could be behind this horrible deception of ruination? The only one who could be behind this most foul plan is the one who uttered the challenge against man in the beginning.

… **"Was he (Adam) worthy of this (the Angels prostrated before Adam) that You have <u>exalted him over me</u>? If You give me respite up to the Day of Resurrection, I will <u>uproot the whole of his progeny</u>; there shall only be a few of them who will be able to save themselves from me." Quran (17: 62)**

Did man think that while Adam – the father of the human race – was under attack by a stealthy, vicious enemy that his offspring would be left alone in peace? So before the law (law of retribution) and before the prophets, the Threatener and the Threatened squared off in a momentous battle. While the Threatener knew much wisdom and devised a heinous plan, the Threatened showed weakness and gullibility. Unable to foresee that he would get nothing for his labors except what was already pre-planned from the beginning, the foolish, false advisor for mankind and his devilish crew embarked on a straight and narrow path already laid out for him.

And he (Satan) swore to them both (Adam and Eve) that he was to them their sincere adviser and well-wisher. So by <u>deceit</u> he brought about their fall… Quran (7: 21-22)

The fool became only a tool in the great scheme of things and races with his band of mischief making followers (men and Jinn) towards the everlasting torment of anguish. In truth, neither he nor his inept crew can take one soul not destined for the Hell-Fire into the Hell-Fire. If one talks about powerless, that is really powerless.

However, the souls being already warned and given

wisdom, insight and power had the ability to be guided. Most importantly, like their father Adam, each soul could fall but then turn back towards the Light in seeking His Forgiveness, Grace and Mercy. Or, like true vicegerents they once were, they could throw away this great bounty of Forgiveness, Grace and Mercy and continue to revel in self-destruction on any level that they would find themselves on. Revel in self-destruction unwittingly until they got caught out. This is easily seen when we see a person doing uncouth acts as if they would be totally free but then become alarmed when they get caught out. Depending upon the deed, the punishment may be mild for a very minor infraction to a multiple life sentence to be spent in prison or even death.

Then do We abase him (to be) the lowest of the low. Quran (95: 5)

So the choice became theirs and theirs alone and then when the curtain closed on their little adventure (for they could not control time) they were cut off and left humbled to the very essence of their existence.

"You have met this fate because you rejoiced on the earth in things other than the truth, and then exulted in it." Quran (40: 75)

THE DARK PERIOD

The world is complex. It is moving in many streams and people are overwhelmed with a flood of knowledge. Science is king and people tend to be either burdened with the pangs of livelihood or numbed by the over sensory input of daily life.

Before I get into the Six Ages of man theory, there are a few things that I would like to clear up. One of the thoughts is the truth that Jesus the son of Mary was not crucified but instead of him being on the physical cross, a man by the name of Simon the Cyrenian was crucified in his place. The historical story goes something like this.

If one has a chance to investigate certain realities, one will find certain common threads running throughout history. These trends have been shown in the Quran as well as hadith literature. One trend that is well observable in the past between various groups of people is the fine individuals who were directly associated with their prophets and who were God-fearing. After that, came a generation not so fine and God-fearing and after that even less quality people. Not only that, but also there appeared among the good folk, from the beginning to the end, people whose twisted minds always seemed to strive against truth, justice, morality and the Fitra of man's innate common sense. This provides one key for our understanding.

The direct disciples of Jesus (pbh) were, after their instructions and lessons, the best of men alive on the planet at that time. The exception of course was Judas Iscariot who was discussed in several of my books but he is of no consequence here. The direct disciples of Jesus had their own disciples and these disciples eventually had their own disciples. The same concept is found in Islam whereby the Companions of the Prophet are followed by those closely associated with these Companions followed by the next generation and so on and so forth.

The trouble here is that the disciples of Jesus were living in a period of dynamic change and turmoil as well as being spread out over lengthy distances. They had no phone or fax to communicate with and their lives tended to be in danger most of the time. The Romans destroyed their base of power, Jerusalem, in around 70 A.D. Persecuted, hounded and rebuffed, these heroes of truth could not survive in a perverted world and so many of them became martyrs. Church history does not speak much about them and for good reason. They have become the great unknowns.

One of these individuals whom we know basically nothing about is Glaucias the disciple of Peter. Glaucias told the truth about the crucifixion to one of his 'listeners'. It was Glaucias who received from Peter the knowledge that Jesus was not crucified. That is, he was not put to death on the cross but that Simon the Cyrenian was crucified in his place. One of his so-called adherents was a man called Basilides who received bits of interesting information concerning spiritualism. Yes, they were interesting but soon they were to become destructive.

This event in itself proves a lot. Basilides had to be just more than a passerby if he was to catch the news of the non-crucifixion of Jesus. In fact, getting spiritual information, which he later used to cause destruction shows that he must have been associated in some inner circle to receive this so-called knowledge but that by his own actions he deceived himself.

Glaucias like you and I can't read the hearts of people and so how was he to know what Basilides was going to do? He couldn't and therefore should not be considered blameworthy for Basilides' actions. Allah alone knows for sure. Jesus (pbh) did warn his disciples as a commandment not to put pearls on the necks of swine because the knowledge one may give may end up coming back to bite the hand that gave it – found in Matthew (7: 6)! In other words, teach the practical and leave the high flying mystical alone although knowledge can be past on for certain reasons and just causes.

Basilides seems to have had a penchant for thinking the life and actions of Jesus were trivial but that mysticism was a hot topic and that pure unadulterated non-scriptural based conjecture was a fun and great thing especially when it could be mixed with pagan religious viewpoints. A man that unstable would be expected to institute a new brand of Christianity. He did. What is worse than that is what his son did with that odd brand of heretical Christianity. His son, not content with twisted Christianity, went all out and developed a kind of pagan-Christian-Gnostic jumble.

So one can see what damage can happen when the advice of the prophet Jesus (pbh) was put in a box and buried. One can also sense from this story the knowledge that Jesus was not put to death on the cross but that another was put in his place does have historical backing. Just because of the wrongs committed by some does not give a person a license to cancel out history. Could people cancel out Jesus because of the treachery of Judas?

This research goes further to show that historical Christianity does have several decades of just about total darkness by which the truer records have been either destroyed or have been lost. In fact, over 50 years of solid evidence has been practically erased. It is true that various chronicles of the early 'preachers of truth' were attempted to be pieced together by certain church historians but what church historians have come up with was woefully scanty and vastly incomplete.

So how do we know about Basilides? He and his son and their system of Christianity made a lot of noise and that noise had to be refuted as it was a grave danger to the very early-organized Church of Rome. The eventual idea of a non-flesh-like Jesus was refuted by doctrine in answer to excessive spiritualism (invention) in some of the epistles found in the Bible today but the history behind it was not explained.

When we come to the fourth generation of the followers of Jesus, a man we don't know much about stands out. His name is Polycarp. Some say that he should be credited with meeting a disciple but in all reality he probably only met very briefly with a third generation disciple. That is, a person who was a disciple to a disciple to one of the original disciples. Polycarp was a good man and the most noteworthy thing about him is his defense of the knowledge of the original Passover date celebrated by Jesus and his immediate followers. That date was the Jewish date of observance but already that date was being challenged by the Roman authorities of the fledgling church such that they felt no guilt about changing one of the original practices of their spiritual leader – Jesus.

Changing the original date of the Passover would be analogous to the Muslims changing the date of Hijra in the

Islamic system. Such a thing would be considered evil in Islam. The idea here is to see what is actually taking place under the vestiges of Roman domination and the piecemeal destruction of the Sunnah of Jesus.

About the time of Polycarp another dark period sets in and lasts a few decades. This period was later filled in by church historians as to what they either wanted to believe or to what would go over smoothly.

The importance of the lessons to be learned here is the fact that New Testament records have been reworked and touched up several times. It is interesting to conjecture that the works of Paul are not totally Paul's writings but consist of what might appear as Paul's philosophy written down by his various disciples. That does not imply that the ideas generated inside these various epistles are wrong. It is possible that these sentiments are correct as well as being clipped and filtered in someway to mask over missing elements. To be sure, his so-called epistles have also been reworked and touched up to fit the ideas of the day but certainly enough of his concepts have been left in to justify someone saying, "This belongs to Paul and his circle." The background and lengthy discussions on various issues seem to be missing in several places.

The evidence of these thoughts can be gleaned from the latest Christian research that is upfront and not from individuals who avidly seek in trying to cover up the truth in dealing with polemics. The interested reader can go on the Internet and get the same information. The purpose behind all of these things is to open up avenues of honesty and truth to people and for those who are interested to let the spirit move them towards a closer understanding of the Oneness of God, the oneness of the brotherhood of prophets, the behavior of the faithful and the oneness of the Way set forth from the beginning.

It has been suggested by some that the Old Testament and

the New Testament have very little value inside of them for those belonging to the various sects called in today's world – Jews and Christians. This is completely wrong and shown by Sayyid Maududi to be erroneous. In fact, the Quran points out this fallacy in the following verse.

O you, to whom the Book had been given! Believe in the Book We have sent down now, confirming the Book <u>you already have with you</u>; believe in it before We distort faces and set them backwards, or lay down Our curse on them as We laid Our curse on the Sabbath-breakers – for the decision of Allah must be carried out. Quran (4: 47)

It is foolish to believe that Allah tells the People of the Book that they can reach the truth from books that no longer exist or that they cannot find. For inside of every human lies that Fitra or state of natural inclination towards the awareness of the Creator Lord which can be, with His permission, the guiding beacon beckoning to the truth, if He Wills, and if the man will take time to use his <u>common sense</u> under the knowledge that there is that Day of Accountability to be faced.

He it is Who has made you vicegerents in the earth. Now whoever disbelieves shall himself bear the burden of disbelief... Quran (35: 39)

Where are we now? Where are we going? How do we get there? The answers to these questions can be simple or as complex as we want them to be. Simplicity may be better than complexity if a person can focus on only a few basic realities. For those of the Jewish faith it would be extremely helpful to concentrate on Deut. (18: 18-19) and for those who call themselves Christians it would be better to focus on John (16: 7-14).

The next idea is to allow that the relationship between the pagan Arabs of old and modern man is not as diverse as it might appear to be. Most of the pagan Arabs believed in

Allah and the afterlife. They had divergent beliefs to be sure (most of them contradictory) while even some believed in the transmigration of souls. These cooked up philosophies together with the idea of tossing the doggie a bone smacked against the real realities being presented to them by the Prophet (pbh). When the worldly man eats all the goodness of meat from the bone and then tosses it to a dog with hardly any scraps on it, is the idea of tossing an 'inferior' being a great big present by which the 'lower' life form should be thankful. This idea shows that man rules but 'their' god but serves which is totally and unconditionally nonsense.

In our every day life the notions of the Creator Lord seem to come to the forefront only on special occasions as if to say, "O, yeah, I do remember You and You are great somehow. Now let me get on with my life." Or one can believe like some of those pagan Arabs believed that the Eternal Being is important enough to invent things about willy-nilly without giving much care or thought about Him and our efforts should be enough to appease Him and all of the other invented nonsense that we have created to protect ourselves.

The idea of following a deep concept of the One God together with the concept of the one brotherhood of prophets was somewhat alien to those pagans but the real blow to their pride was the concept of the Resurrection and all that the Resurrection implied. It implied accountability and guidance of the self to do one simple thing. That simple thing was to submit to their Creator Lord not only in recognition, which was simple, but in trust, thankfulness and obedience to His Messenger which turned out not to be so simple when they still wanted to hang onto crass beliefs. It would take not a symbolic 'toss the doggie a bone' trust or a put Him on the backburner trust or a temporary trust but a continuous, constant love, fear and optimistic trust.

Modern society is very self-centered and materialistic as

well as sensory such that an overload of sensory input actually dulls the senses of inner contemplation. Nevertheless, science is king and will show the way. Or will it be able to show the way? I think science might be helpful if utilized in the correct way but for showing the way, I personally don't think so. Science is a tool and no more than that and it can be used like a drug to forget about the real nature of man thus causing the careless to stumble.

It was stated in a previous book, *The Real Holy Grail: The Messiah on Trial* that there are Six Ages of man. In that book certain things were detailed about that theory. According to that theory if the ages of man have been getting more materialistic, then what about the Fifth Age? Shouldn't that be more materialistic than the Fourth Age? The answer is no. It is not because of the ages of man ended for there is a Fifth Age and a Sixth Age to come. It is because the Fourth Age represents <u>an end of an era and a reward/punishment concept before the Resurrection</u>. After all, the Fourth Age is the age when prophethood stopped. It is the age when the Sun will reverse its course in the sky and the 'beast of the earth' will come forward.

Now these concepts can be disputed as just wishful thinking. Mankind was told to ponder over certain events. Mankind was told to study the records (holy writ) as they should be studied. Mankind was told to contemplate about the things around him and his relationship to his Creator Lord. Mankind was also told to contend with himself and be thankful to the One Who gave him life and the One Who will return him to life after death.

Say: "It is He Who has created you, and made for you the faculties of hearing, seeing and understanding: but you are seldom grateful." Quran (67: 23)

The idea of how mankind will be raised up has always caused a stir one way or another and facing the consequences of one's actions which implies a test of not one's beliefs so much as the idea of right guidance. That sends shivers down the spine of so many of the 'faithful' Why?

A person, for example, who has a strong faith in cannibalism and practices it daily, expects his reward. What kind of reward can that misguided soul get for his long labors in faith? He should not receive a good reward because he misused faith for his own purposes and his good qualities were rendered useless for the Hereafter. The Hereafter has rules and these rules are not broken for anyone. So it behooves a person to seek out and find those rules because mankind does not invent the system of salvation. Hence, the concept of the brotherhood of prophets and what that means as to the true application of man's behavior should become apparent. That behavior can be illustrated from the following hadith. **Abu Huraira reported God's Messenger as saying, "Between the two blasts there are forty (and when Abu Huraira was asked if the forty referred to days he refused to say; when asked if it referred to forty months he refused to say; and when asked if it referred to forty years he refused to say). God will then send water down from the sky and they will sprout like vegetables. The only thing in a man which does not decay is one bone, the tail-bone, from which the whole frame will be reconstituted on the day of resurrection." (Bukhari and Muslim) In a version by Muslim he said, "Every son of Adam will be devoured by the earth with the exception of the tail-bone from which he was created and from which he will be reconstituted."**

The connection between this hadith and man's expected behavior may seem tenuous at first because it seemingly is so disjointed from the topic at hand. That would be natural if one is not thinking of joining up the connections. Admittedly, that is very difficult to do when one is not engaged in that activity. However, its startling applications concerning what was written about the Fifth Age of mankind in the previous book is worthy of note.

Some people may look upon the above hadith and argue scientifically that it is absurd. Some have argued that this hadith is a reality by various means.

Scientifically, it can be argued that the tailbone of very ancient man going back over one hundred thousand years has been consumed and that the above hadith is either forged or is an invention – God forbid. Some people can counter this position by declaring that because matter cannot be destroyed – it can change form – then reconstructing the tailbone is within easy grasp of the Creator Lord. If the atoms of this tailbone have been scattered for tens of thousands of years all over the world, then the Creator will have no problem to originate His creation of man. That is true! Even by making outlandish examples of these scattered atoms like one atom of the tailbone getting attached to a rocket, which carries it to the Moon or beyond the solar system, would pose no threat to having it being collected together with its remaining parts to form the tailbone from whence the man can be put together. Such a thing would be easy for the Creator Lord.

However, why go to such lengths of the unusual when another yet more practical answer seems to be right before us? An answer that is simple and scientific. But answers are not so simple to grasp and therefore the discussion must be somewhat complex to make the complex turn into the simple. So this discussion will need people to show a lot of patience.

First of all, the above hadith does not go into the physics of alternating states of matter of the tailbone of mankind. It says that man will be reconstituted from the tailbone and that is an actual fact. The Six Ages of man theory can explain this. In this theory the Fifth Age of man, the reality of the Messiah and the kingdom given to him are close at hand. So from one hadith the conclusion can be made as to what final road man needs to travel on for success and what final goal will be given to those who desire success in the Hereafter. The answer to that becomes relatively simple. It is Islam.

Although Daniel in the Old Testament and Paul and Peter in the New Testament touch lightly on the future of the human race, it is very difficult to gather one's thoughts about these issues. First of all, it was not their primary job to explain these things. Secondly, what is presented is not so comprehensive enough to put an order to it. The most important verse in the New Testament comes from the Second Epistle of Peter whereby he talks of the heavens on fire and the Earth melting and then he goes on to declare that nevertheless there is going to be a new heavens and a new Earth – 2Peter (3: 10-13).

However, by going to John (16: 7-13) one does get the instruction of what to do and where to go and the person preordained to give those answers – Prophet Muhammad (pbh). (See appendix C).

Another concept that is related to the tailbone hadith is the second death/last death idea. The concept of the last death is given in several hadiths one of which is found below.

On the day when all kingdoms will perish, according to a tradition, the holy Prophet said, "Allah will take the heavens in one hand and the Earth in the other, and will declare, "I am the Sovereign: I am the Ruler. Where are the other rulers of the Earth? Where are those tyrants? Where are the arrogant people?" (Ahmad, Bukhari and Muslim)

The meaning of the last death is that all things except Allah are to die and that what is brought forth is the commandment that there shall be no more death to the rejoicing of some and the utter horror to others who are trapped by their own devices in the Hell-Fire. That a new creation will be brought forth cannot be doubted such that man (those who are successful) in his wonder and praise for his Lord will see and experience things never before seen or experienced.

When a person dies the normal death he doesn't really die. Of course, his body-like shell (the outer body left behind as a lifeless husk) decays. The exception to that are the bodies of the prophets that remain in tact without the hint of decay. But man's second death hasn't happened yet although many might like it to be the case.

What is this second death? By looking at the Quran and hadith material, one can see that every nation and every tribe and every person has an appointed term that can neither be put forward nor delayed. It is a time of unknown length that is given to those things mentioned and when the time is over, the groups involved in this time are locked up (put as in sleep) so that others may enjoy the fruits of their leanings and the lessons to be learned and the tests to be undergone on this Earth. As to what exactly this term of appointment is cannot be known as it is in the Unseen.

No people can hasten their term, nor can they delay (it).
Quran (23: 43)

But what can be known is that every fruit has its own peculiarity and growing conditions. Tomatoes, grapes, melons, apples and peaches are all fruits that can be harvested in a season. Yet these fruits do not grow at the same exact time, under the same exact soil conditions and in the same manner. So how could people know what can't be known except by the Creator Lord? Souls like nations enter the earthly domain in surges. What did one think? The idea that all souls are born into the Earth at one and the same time would be bizarre.

The Quran gives us not only the stages of mankind as a fetus till birth but also stages of man during his life and other things as well. Nations like people come like the surges of the ocean's waves – one after the other and last as long as their term has been decreed for them. Then other nations and peoples come into the earthly sphere to be tested as to their values and obedience to their Creator Lord or lack thereof.

This is not the second death. This is their trial and all have to undergo their trial or there would be no justice from the One Who is called The Most Just. This stage could be called the period of the prophets whereby each nation called an Ummah has recourse to that which is laid down for what man calls salvation. In other words, mankind was promised that he would be given an opportunity to show his inner being and fealty to the Creator Lord for his creation.

THE GIFT OF ALLAH TO MAN

WARNING! What is going to be presented in this chapter is a radical idea concerning the Six Ages of man theory. This idea does have scriptural backing as well as backing from hadith literature. It also shows a strong connection between the monotheistic religions and the deeper understanding of the word 'Muslim' in all ages as it pertains to the oneness of the brotherhood of prophets and the Unity of the Godhead. It is a concept of unity of belief, which marks a person off not through shallowness of belief but of steadfastness in this unity of belief culminating in that religion known as Islam.

Any theory that contradicts the integrity of the Quran and Allah's Messenger is wrong. The Six Ages of man theory will not appeal to man's ego as to being of a special race, named religion or social position or stature of race. It is a theory that strictly depends on the Unity of the Godhead and the oneness of the brotherhood of prophets down through the ages culminating in the last revealed Divine Message and <u>the Messenger of the Messengers</u>.

Therefore, it goes without saying that it is somewhat of a complex theory that takes a great deal of understanding and patience to perceive. Since it is a theory opposed to a special race or class of people, it will not go over well with those who follow that line of thought because it takes away their pretence of being chosen. Man cannot choose to be one of the 'chosen ones' simply by cleverly twisting Holy Scripture in perverse ways to uplift their own souls nor can man come to a common agreement to negate truth and therefore create an imaginary paradise for their own habitation.

To understand this theory one needs to understand the Oneness of God and the universal oneness of the brotherhood of prophets and the oneness of the term 'Muslim' as it is used in its <u>generic</u> sense without time or border constraints. Above all, one needs to utilize patience in understanding the complexities of this theory as they are being unfolded.

Any theory can be challenged as to its legitimacy as well as being supported by facts. It is how one applies those facts that determine, in part, an understanding of the problem at hand. For example, the prophets did not go around contradicting themselves nor did they contradict each other. With this in mind it is hardly fitting to start making contradictions where none exist except in the fertile minds of those who seek contradictions. And those who seek contradictions do so for an apparent reason. That reason is pride or neglect so that they can put themselves above others in a matter of salvation or even to deny salvation. Salvation does not occur by magic and it does not occur through man's independent goodness. Salvation occurs through His Will alone by His Grace, Mercy and Forgiveness.

Now we are getting closer to approaching a unified understanding of the second death and the tailbone hadith and eventually that awesome and terrifying period called the Last Day of Judgment. As a scientist might use forceps to turn the pages of a very old manuscript instead of a hammer and a crowbar, one must look at the reality of the situation first before charging ahead. The reality has always been that people who were staunch at following the Unity of the Godhead and the oneness of the brotherhood of prophets were called Muslims and that was in <u>any age</u>. It was mainly the builders of society – the priest class – that caused irreparable damage of sectarian differences.

The same religion has He established for you as that which He enjoined on Noah – the which We have sent by inspiration to you – and that which We enjoined on Abraham, Moses, and Jesus: Namely, that you should remain <u>steadfast in religion</u>, and <u>make no divisions therein</u>: To those who worship other things than Allah, hard is the (way) to which you call them. Allah chooses to Himself those whom He pleases, and guides to Himself those who turn (to Him). Quran (42: 13)

In truth this Ummah of yours is a single Ummah (one community) and I am your Lord: so worship Me alone. But the people (of their own accord) cut asunder their (one) Creed <u>into many religions</u> yet they will have to return to Us. Quran (21: 92-93)

These differences, instead of elevating man, debased him and caused disharmony and disunity. It could only have been the priest class who started the ball rolling, as they were the ones who controlled the records from of old in the first place. Nevertheless, people who found certain perversions in their own soul to be enchanting did not mind following corruption.

Say: "Everyone acts according to his own disposition: but your Lord knows best who it is that is best guided on the Way." Quran (17: 84)

It has been claimed in the New Testament that the souls

asleep in Christ shall be raised up that is that they will be partakers of the new age to come. Who are these people who have fallen asleep in the Messiah-soul and what relationship do they have with Islam, the Fifth Age and the second death and ultimately with the tailbone hadith? Are we led to believe that contrary to the laws of true religion that only a certain race, country or creed will be raised up? If so, that does not bode well for the universal truth. To answer that question one has to look at the various stages of the future life and the gift of Allah to His servants. Only by diligent research and by study can these things become clearer to the people. It is unfortunate that many people do not know these things even many Muslims so it behooves them to test these sentiments if they are in doubt. What can be known in general terms is that the following things will occur – Allah knows best!

(1) End of the Fourth Age – The world's population has been drastically reduced. The world is about to enter its Fifth Age and so the world by the command of its Lord has undergone radical changes to prepare the kingdom of heaven on Earth or that what is known as the era of the Messiah. The defilers of truth, the hypocrites and the disbelievers have been removed along with the hideous Gog and Magog and their related friends. Even before that, the Anti-Christ has been destroyed. So the good people who have died previously will be raised up according to their intentions as to when they are to be raised up and brought forth to participate in this earthly paradise as children. Children? What did one expect? Did one believe that one would be walking around (except the most blessed) as a full grown man when Jesus himself has children by way of marriage and that the Sunnah blesses people for having children? How else should man come into the world but as a child? However, that

is only for those who are to be blessed.
Until when death comes to one of them, he says: "O my Lord!
send me back to life,
In order that I may work righteousness in the things I
neglected." By no means (will he be allowed to do that)! It is
but a word he says. Before (these people) is a Barrier until the
Day they will be raised up. Quran (23: 99-100)

(2) The life of Jesus again asserts itself. He will get married and have children and before his death all people will be in a state of harmony of one belief in accordance with Divine Law and Promise.

(3) This kingdom is on the Earth only and that kingdom will follow the earthly law of birth and death (the second death promised mankind) such that 'entropy' will still be found to exist as well as gravity and the other laws, called by man, the so-called laws of nature. Now this kingdom of harmony is supposed to last 1,000 years after Jesus dies (Allah alone knows best). That is, the harmony and life style will be in agreement as to the knowledge of the pure Book (Quran) and the Sunnah of Prophet Muhammad (pbh). The devils have been imprisoned in their place of containment so that the heaven on Earth promised by Allah will proceed and mankind will be grateful to their Lord.

(4) After this period the devils are unchained and mankind again grows diverse and shows various states of ingratitude for the Earth pattern must hold true to its course and a state of entropy takes its natural disposition. That is that what was shown by every nation as to the generations born during a prophet's life and then the next generation and then the next generation and so on and so forth. Then when things get so diverse and so ungrateful toward the Creator Lord, a gentle breeze will be sent to take out any

human who is left alive that has even only a mustard seed of goodness left in his heart.

(5) Now begins the very short Sixth Age. This is the age of vileness whereby only the rubbish amongst mankind is left on Earth.

A close study of scripture can reveal these things. But who are these people who are asleep in the Messiah? Obviously by looking at Daniel (12: 10-13) there are people who will still be alive after the carnage of the Anti-Christ and the Gog and Magog. Living people get married according to the Sunnah of the day and people are known to come into the world as babies by the regular method and not by magic. So who are these people?

These people are souls who are highly blessed especially to be born at that time of honor and righteousness. So who are those people? These people have right guidance placed in their souls as a favor from their Lord and so who are these people?

If one understands the words of Jesus and the words of Paul we can now understand who these people are who receive the gift of Allah and rejoice in His absolute Oneness, Mercy and Forgiveness. Yes, it is true that they were asleep in Christ but who are these people? We can know who these people are by several methods and one method is by knowing who these people are not.

Can one imagine a soul being born in that blessed era by proclaiming the following? "O You unfortunate believers how you struggled in Allah's way and how you suffered hardships for His sake and toiled throughout the ages but look at us and behold. We neither suffered nor toiled. We get the best of both worlds. Without work or toil, we get a free ticket to enter Paradise because of no effort on our part and no test of our faith. We are a chosen people of do nothings and getting all the choicest places." That of course is absolute nonsense! It also belies the saying that they are asleep in Christ the Messiah as well as canceling out Holy Scripture and various hadiths.

So who are these people? Real religion does not proclaim a chosen race of people nor does it proclaim a lottery lucky draw. It most definitely doesn't proclaim a testimony from people whose tongues wag in praise of Allah but whose hearts have another agenda. So who are those people?

A close look at the Messiah tells us who they are. They are people of <u>any era</u> who really and truthfully believed in the Oneness of God, the oneness of the brotherhood of the prophets and those who learned and practiced true faith and followed their prophets in trust and sincerity. How do we know this?

Jesus proclaimed of himself who he was and how the builders (of society) rejected him as the cornerstone but how in spite of their rejection, Allah made him head of the corner Luke (20: 17). Paul described him as being the first of the fruits raised up or chosen – in this case for a particular job. I. Cor. (15: 20) and I. Cor. (15: 23). Allah plants the fruits and He is the harvester or husbandman of those fruits as described in the New Testament. The fruits are plucked by Him and so harvested. They do not harvest themselves and so Jesus in his own words is that chosen vine John (15: 1-2).

But Jesus declared his special gift – a gift given to him but not to you or me. This gift is one whereby he could enter and leave the world not by his command but by his Lord's Command John (10: 17-18). He also proclaimed that he had other flock (people) to whom he came John (10: 16). The son of man did not lie nor was he deluded but he spoke the truth.

Did the son of man born in the Third Age in the time of Noah (pbh) and known as Melchizedek lie? No, he didn't. Thousands of years past and he came down to the Earth and lived a normal life and spoke with Abraham (pbh) and even ate and drank with him. This means that he was no spirit but a living human being. Does that make him God or God's son? God forbid!

Did the son of man Joshua bin Nun lie? No, he didn't. First he was a servant of Allah and then he was a servant of Moses (pbh). It was not imputed to Moses to lead the Children of Israel into the Promised Land. It was imputed to Joshua the son of man. Here are some of Joshua's words when he was a servant of Moses. This concerns the enlightenment Moses (pbh) received from his Lord concerning an understanding of some things hidden from normal eyes. This discussion takes place on the quest of Moses to seek out one of Allah's servants who had been gifted with special knowledge.

He (Joshua) replied: "Did you see what happened when we took ourselves to the rock? I did indeed forget (about) the fish: <u>none but Satan made me forget</u> to tell (you) about it. It took its course through the sea in a marvelous way!" Quran (18: 63)

Some other words of Joshua (pbh) taken from the Old Testament concerning his warning to the Children of Israel are as follows.

Now therefore fear the Lord, and serve Him in sincerity and in truth: and put away the gods which your fathers served on the other side of the flood, and in Egypt; and serve you the Lord.

And if it seem evil unto you to serve the Lord, choose you this day whom you will serve; whether the gods which your fathers served that were on the other side of the flood, or the gods of the Amorites, in whose land you dwell: but as for me and my house, <u>we will serve the Lord</u>. Joshua (24: 14-15)

Does that make Joshua a God or God's son? God forbid!

Then there is Ezekiel (pbh) who is even called the son of man in the Old Testament. He was a true servant of the Creator Lord. He was given special knowledge of the future and past and he received a dark book containing mysteries. Does that make Ezekiel a God or God's son? God forbid!

Then there is Jesus (pbh) who was such a tireless worker and suffering servant in patience. Some thought him to be a madman or sorcerer but turning out, to the embarrassment of the disbelieving folk, the chief cornerstone of the edifice of true faith on Earth. He was one of those beautiful souls amongst the first fruits raised up and harvested by his Lord and came into his existence in Allah's Perfect Plan by a simple command of the word 'Be'. Ever the worthy servant, he is the son of man right blessed in this world and the world to come. All hail to the conquering servant who conquered sin in all of its forms and temptations to be the pure chalice that holds and binds together all the sincere servants of the One God for whom he was appointed. Hence, he becomes that one whose authority shall usher in the kingdom of righteousness and the religion of truth – not to one nation but to the <u>whole world</u> where harmony and unity of purpose shall reign supreme. These people will be God's chosen delights amongst man. Does that make Jesus a God or God's son? God forbid!

So who are those people of the tailbone or second death? As people are born into the world of the new age – the Fifth Age or the kingdom of the Messiah – they shall, like Jesus, have many children. They will be in a blessed time or a garden on the Earth (paradise) but they shall be on the Earth (a new and blessed Earth) but an Earth that still has its dominions and its dominions have rules. Therefore, people who are born must <u>die</u> and not live forever on the planet, as Jesus must die. Their bodies will decay except for the tailbone and from that tailbone they shall be reconstituted and brought forth to the Judgment. Some will be given their records in their right hand and some will be given their records in their left hand on the day whose length shall be lengthened when the oceans explode and the mountains become like dust.

That is another story altogether and the Quran gives us a very vivid picture of those events. However, the Earth is the Earth and so a pattern must be followed. Those who are born during the life of the returned Jesus (pbh) are higher in quality than those born at a later time. And those born at a later time will be more excellent than those born after them and so on and so forth as to their future rankings in Paradise.

The Word changes not with Me and I do not the least injustice to My servants. Quran (50: 29)

Then there is the vilest of periods called the Sixth Age. According to the Six Ages of man theory each age of man is much shorter than the previous age. This shows that after the blessed event of the return of Jesus, thousands (Allah knows best) of years will pass but not tens of thousands of years when the Fifth Age will come to a close by a gentle breeze sent by Allah to collect any soul with a mustard grain of goodness in it. Therefore, the idea of the tailbone hadith not only becomes clear but also is scientifically more feasible without all that philosophical rot being thrown in to confuse the issue.

Christ being the head cornerstone has several meanings. The first thing is that the pattern from the Earth was born in him as was expressed in one of my books with the section heading *'From the Earth the Pattern Comes'*. This shows the development of the religious movement through him and the oneness of the brotherhood of prophets.

The next thing is that the Messiah is the chosen one to lead this kingdom of heaven on Earth and this adherence to the one brotherhood of prophets is the base on which a person is justified in being classed as an earnest believer.

So the Messiah-soul is not about just Jesus but about the one brotherhood of prophets that he represents and therefore the people who are asleep in Christ are all of those who were earnest in fearing and loving the One God and following the righteousness of the message delivered unto them from the various prophets of old. That of and by itself pulls the plug on the smug and self-righteous people who think that salvation is for them only as well as the people who have tried to manipulate religious thinking for their own self-glorified benefits.

The builders of society who rejected the Christ are not engineers or plumbers. They are they who were supposed to build society as to the most precious thing a man can pass on to the generations but because of their corruption, others were led into corruption. Without the law of the Sunnah there would be chaos and divergence and as the Quran shows there would be destruction. Destruction comes by thankless, perverted behavior in refusing to submit to God's Plan. Who else would the builders be but those who are of the priest class whose job was to order the era of right guidance and belief?

Also responsible as builders are those who thumbed their noses against right guidance and went their devious ways because they refused to accept gratitude for their creation and wanted to flaunt their 'freedom' in unscrupulous behavior because it was more fun to be attracted by worldly delights rather than burden one's soul with following Divine Guidance.

There are three things that separate man from the Hell-Fire. The first is the awareness of the One God. This awareness is not philosophical or mystical. It is what has been described from of old. What has been described from of old? It is that man must subdue himself according to the wisdom of the blessed Messengers sent to them. That is, to follow and obey the Messenger without qualms or hidden agendas and to pay them strict and willing obedience. That means to avoid what was told them to avoid and to perform the acts that they were told to perform and do these things with a cheerful heart.

The second thing is to cultivate the internal justice that they have received from their Messengers and look towards that system as a guide to bring equality of behavior to man. The Earth subdues the physical-materialistic man but the soul/spirit can subdue this process through establishing justice amongst themselves thereby bringing order, balance, harmony, unity and justice to the Earth rather than injustice and hypocrisy. Hence, justice will be meted out on the Last Day fairly without corruption and in equality to those who have lived on Earth.

The third thing is to call to mind the Creator Lord in all activities engaged by man to remember Him often as to His Grace, Mercy and all encompassing Forgiveness – not as a toy-thing but as a Reality more real than life itself. His Beingness is more real than ours and so why not remember Him as instructed. As instructed? That is where His Quran and His Prophet's Sunnah come in. It is a roadmap to success. Success not in building things that will turn to dust – for build we must – because these things are only secondary. The real importance in life, which is primary, is the remembrance of Him Who created us. That is where true guidance comes in and true guidance was given by the brotherhood of prophets and no one else.

And whoever repents and does good has truly turned to Allah in repentance. Quran (25: 71)

And Allah also enjoins: "This way alone is My Right Way. Therefore, you should follow this Way and should not follow other ways lest they lead you astray from His Way... Quran (6: 153)

People might consider this to be a highly philosophical discussion. However, many people should realize that throughout the ages they have been blessed with certain scripture that claims to lead one to eternal happiness or the other place of ill repute. Since man has those records in one form or another, it is up to him to utilize them as he was told to utilize them and not make sport with them. After all he is only building a record for himself and he will live and die and eventually live again on his own record. That for sure is a very intriguing thought.

As to the people who have died before this time, there is no need to worry about them. The Creator Lord is All-Encompassing and He knows His Own system better than anyone can imagine. It is up to the members of society to find their way to spiritual satisfaction and this can be done through understanding and making a commitment to the truth. So to the Jews let them study Deut. (18: 18-19) and to those who call themselves Christians let them study John (16: 7-13) and let them understand these things well.

The theory of the Six Ages of man is just a concept put forth to explain man's evolution in part as to groups or communities being sent forth in stages to live and experience a trial on Earth. It only presents a workable thesis to explain in part the dynamics of a system. In this system true man does not come from the entrance of souls as spirit forms but through the acceptable system of the mother becoming pregnant and then giving birth. Any other system would be abusive and therefore unacceptable as to salvation. So salvation is through the approved line for all must submit to it.

Secondly, in the Third Age the individual known as Melchizedek was shown to have parents according to the Dead Sea Scrolls and he was born in the era of Noah (pbh). Noah who lived around thirty some thousand years ago came from a community that had people living for hundreds of years. Noah was no exception and neither was his community or else a strange thing would have happened. The people of Noah for the most part rejected him because of his insistence to the truth of the Day of Accountability and because he was a commoner like them. If he lived hundreds of years more than them and their children and children's children, then he would be accounted for being more than a man. This shows that the prophets fit into their societies perfectly and that the accusations against them of being only men, and non-ostentatious ones, confused the boisterous and arrogant from among their tribes.

And We delivered him (Noah) and his people from the great Calamity,
And made his progeny to endure (on the earth). Quran (37: 76-77)

Thirdly, the line of blessedness was preserved in the Ark as it survived the catastrophic flood. This is from whence the line of true humanity sprung by which 'others' would receive the gift of true knowledge surviving the flood in other parts of the world but as to their precise condition it is not known. Anyway, the family of Imran was allowed to survive and that is the important thing.

Fourthly, the concept of Prophet Muhammad (pbh) being a prophet to the whole world and as to being the leader of the prophets in that he is the leader of the Way of the brotherhood of prophets is enhanced and brought more into focus in the concept of the Fifth Age by which souls are rewarded in proportion to their sincerity of beliefs and action upon the same rather than tongue wagging. Also, enforced is the idea of him being given the white and the red treasures and the tailbone hadith spoken of earlier becomes more clarified as to its understanding.

In conjunction with this is the commentary by Abdullah Ibn 'Abbas on the topic of the gardens given for the believers. There are many gardens. Two of which are the Eternal Garden which is the reward of finality in the Hereafter and the worldly garden which is the reward of the strivers in faith of any age regardless of former race, creed or religion. The Earthly kingdom or garden was promised to the Messiah and there are several hadiths relating to that heaven on Earth period whereby souls must arise, awaken and come into the world by being born of a woman and not by any other method.

This concept is a reward for the soul to <u>purify itself</u> and a further chance for the soul to be further trained and attuned to the Reality of the Oneness concept of the Creator Lord in preparation for the 'light' judgment to be faced by the true believer and to be raised up on the Last Day and washed in His Spirit as though they have come out from taking a cleansing shower.

It may be stated by some, "What kind of a reward is that?" Those same people might ponder over these hadiths.

According to a reputable hadith a person asked the Prophet about the good deeds that he had done during the days of ignorance and about what reward Allah has given him. The Prophet told him that Allah has rewarded him with Islam.

A hadith Ibn Marduyah related from Yazid ar-Raqashi. He says a person asked the holy Prophet, "We give among our wealth so that we become well-known. Shall we get a reward for this?" The holy Prophet replied, "No." He asked, "What if one has the intention both of Allah's reward and of reputation in the world?" The holy Prophet replied, "Allah does not accept any deed unless it is performed exclusively for His sake."

The concept of a person who dwells on the Hereafter rather than an obstinate person is like the difference between day and night. Never has the Creator Lord left mankind to his own doings without the essence of His Mercy, Grace and Forgiveness being in close abundance to be tapped into. Forcing a man to be intelligent is not the way of the vicegerent-man to be treated. Rather the vicegerent-man should be able to utilize his given intelligence to come to a realization of where and how his bread is buttered if he is able. Modern life is a good example.

In modern life some of mankind has been given

understandings of science to make his life easier and other helping devices like counseling and retirement payments so that he may live a more fruitful life and have a chance to heal his physical and mental wounds. With these life advantages, including servants to assist in drudgery, does mankind remember to show gratitude to his Creator Lord or does he profess other things? These other things might be that he thinks that his own genius creativity and his ability has been the real engine that has evolved into the super species that is responsible for providing these cultural refinements. If so, then the idea of 'peanuts' for God but 'pounds' for him shows just how much gratitude there is given in thankfulness to the Creator Lord.

So the Fourth Age is topped off by one of the religious beliefs of mankind called ISLAM and the conquering of Arabia and Arab lands and the sending of messages by the holy Prophet (pbh) to the worldly empires of that day to accept or reject the pure religious belief sent down for mankind. What was sent down was an Arabic Quran to the Arab nations and then by an evolutionary process, those to whom were to believe. Some did believe such that Islam became a molding force in the world of the Fourth Age.

These actions cannot start in a vacuum and therefore had to start some place with some people. It started with the Arabs in Arabia. It (the movement) had to start on Earth and had to have some physical place and a physical people to start with. Why the Arabs were granted this favor or why Arabia was chosen is up to Him to decide. But it is only common sense that it had to start somewhere and with some peoples. There is nothing magical in that because all human beings have to eat their daily food and this is a 'law' for planet Earth. Therefore, the world religion looked for and discussed by prophet Jesus (pbh) must be from this Earth and not some magical or unreachable thing from the sky.

It is also seen that souls are given a chance to submit to this system of Divine Grace before their term is up and then they must sleep while other 'groups' have a go at finding their way on the one road that leads to success. This also shows that His system is perfectly just and fair and while many things are relative, the system itself (whether we understand only a little of it or hardly anything at all) is really relative to the situations set up for any times, any age and any place. So, essentially that is the Gift of Allah to mankind.

Therefore, the two most important things that the theory of the Six Ages of man can suggest that mankind follow with hope in their hearts are the Quran and the Sunnah of Prophet Muhammad (pbh). Showing the conclusion of the Quran and Sunnah as the relevant pathway to external and internal success has taken four books and a lot of divergent thoughts and expressions. The world is often so complex that a person can get lost in a forest of trees. Calling out of this mess, while giving clear and unambiguous meaning to life, is the Quran and Sunnah of the Prophet. It is no wonder that Jesus the son of Mary will follow the same design placed in the Fourth Age for man to be utilized to its utmost in the Fifth Age of man. That is one reason that the Prophet had declared that the people of that time will be as good as his Companions or even better. This is yet another testimony to the Might and Greatness of the Creator Lord – Allah.

According to this theory, it is mankind who will be arranged in ranks of sincerity based on the quality and devotion put forth in their lives towards the Way. This Way is the obedience shown to the one brotherhood of prophets in any age. So, the souls with the highest sincerity are brought forth and are living during or born during the life of Jesus' second coming and so on and so forth. These souls are born and die and their bodies are returned to the earth – rank upon rank. Then a gentle breeze comes and all people of even the slightest worth will die and their bodies will return to the earth.

The Six Age is also based on rank for there are worthless souls and there are even more worthless souls to come after that. They also shall be returned to the earth – rank upon rank. However, there shall still be people living on this planet when the Sixth Age comes to a crashing end. The end of the Sixth Age will be when one single mighty Shout will issue forth. At that time the Earth will be told to surrender what it has to surrender and mankind (Allah knows best) will be brought forth collectively from their tailbones while the Earth and the heavens will undergo a traumatic and dynamic change such that the Earth will become a flattened plain holding the raised up souls, the sun will be brought close and the heavens will be rolled up. This is the start of the final court of the Day of Reckoning whereby the new body given to the soul is being prepared to enter Eternal Life. So the real Gardens of Paradise contain such bliss that the minds of men have not imagined. These Gardens are far superior to the heaven on Earth and not only because they are forever instead of temporary.

They say: "Shall we really be restored to our former state?"
"What! – when we shall have become rotten bones?"
They say: "It would in that case be a return with loss!"
But it will only need a single shout (compelling cry)
When they will forthwith appear in an open plain. Quran (79: 10-14)

An outer worldly manifestation of religious beliefs only fools the foolhardy and is accounted as practically worthless. This is likened to the shell of a coconut whose meat and juice have been extracted while leaving the husk intact. A buyer of that coconut who cracks the shell expecting to find the tasty meat and juice inside is rewarded with only empty air and all he is left holding are the broken shards of a worthless coconut.

There are some people who might argue that if Allah had really meant the Prophet to be a prophet to the whole world and the Quran and his Sunnah a universal practice among the faithful for the benefit and uplift of mankind, then he should have been sent speaking a universal language that everyone could understand. Therefore, he is a prophet for the Arabs only. That is utter nonsense for several reasons. First of all the people were told about him by their prophets. Second, the priest class covered it up. Third, the people have had the chance to study their religion in a well balanced temperament but if they decided not to do it, then why blame anyone for this failure except the one's who committed it. Finally, the Gift of Allah doesn't exclude those who were faithful to their prophets. Allah is not cruel or hardhearted. He gives His servants a chance to be refined in the Fifth Age and that is the universal age for embracing Islam.

In a bizarre twist of subtleness, the universal religion of Islam becomes the universal religion and way of life for all of mankind born and living in the Fifth Age and the foolish and egocentric man is totally powerless to stop it. On game shows men and women can match wits with their counterparts but trying to match wits with the Creator Lord is absolute lunacy. For it has already been proven to the serious students of religious thinking that the return of Jesus the son of Mary will usher in the new era or that earthly garden of abode having as its basis the Quran and the Sunnah of Prophet Muhammad (pbh). Under the leadership of the Messiah and others, the world will be truly a just and admirable place to live and a place to die and be resurrected from.

Now if mankind were to believe that they would do something about it for themselves. Although there is no compulsion in religion man does have the faculty of commonsense. He can use his commonsense or as the case may be he can throw it away. That is his free choice.

They will further say: "Had we but listened or used our
intelligence, we should not (now) be among the Companions
of the Blazing Fire!" Quran (67: 10)

However, what was said to the Prophet by Allah in the
Quran still holds true in any sense or in any condition. That
is, that he cannot cause the people in their graves to hear and
as Jesus declared, 'Let the dead bury their dead.'

Nor are alike those that are living and those that are dead.
Allah can make any that He wills to hear; but you (O Prophet)
cannot make those to hear who are buried in the graves.
Quran (35: 22)

For those who reject their Lord (and Cherisher) is the
chastisement of Hell: and evil is (such) destination.
When they are cast therein, they will hear the (terrible)
drawing in of its breath even as it blazes forth.
Almost bursting with fury: Every time a Group is cast therein,
its keepers will ask, "Did no Warner come to you?" Quran (67:
6-8)

An interesting understanding is brought forth from
Quranic verse 22 in Surah 35. One can now match this up
with the New Testament version found in Luke (9: 60).

Jesus said to him, "Let the dead bury their dead but you go
and preach the kingdom of God." Luke (9: 60)

It is interesting not only in a comparative religious
viewpoint but also in an understanding viewpoint. Of course
the truth is subjected to the test of time and the harmony and
unity of meaning. It is to be supposed that there are different
levels of meaning for this Quranic verse (35: 22) but let us look
at one of them and see how it equates with the New Testament
version which is also found in Matthew chapter 8 verse 22.
However, Luke's version seems more accurate.

If one takes a physical meaning to these verses, then
things become quite perplexing. Imagine people taking these
things literally. Certainly there would be a terrible uproar

when decaying bodies are left to rot and cause disease because people are waiting for dead people to popup out of their graves to bury other dead people. If Jesus had meant this, even his own disciples would have treated him as a deranged individual. However, it is to Islam that one must turn for the understanding.

It would be fruitless to believe that the Prophet (pbh) would be told to do something useless and imaginary and simply something that would make no real sense whatsoever.

Nor are alike those that are living and those that are dead. Allah can make any that He wills to hear; but you (O Prophet) cannot make those to hear who are buried in the graves. Quran (35: 22)

The Prophet himself gives the reason why this object of making the dead and buried to hear would be fruitless. He declared that when a person dies, all blessings stop for him except for three conditions. These conditions are in general formed from the special prayers of the good and faithful children for the forgiveness of the deceased, honest money that continues to work for the benefit of the believers and the passing on of the right kind of knowledge that in effect can be utilized to uplift and benefit mankind especially in bringing back a remembrance to Him Who is solely responsible for our creation. It is granted that most people do not get a shot at this boon and so the blessings to be added to the scale of justice for the determination of whether the individual will receive his record of life in his right hand or left hand stops at the period of death.

So to talk with a dead and buried person would perform no logical function at all. It would not be to his/her benefit to listen to any such speech. So what are we dealing with here?

In the spiritual sense we are dealing with individuals whose hearts have been cutoff from the truth. These individuals would not respond in sincerity to the truth (maybe by way of fear) even if a whole host of Angels would be allowed to appear before them. They would think that it would only be a bad dream or an illusion. In other words, the light of awareness to faith has been cutoff and for the purpose of creation of his blessed being he remains a dead and useless object – except for making mischief – but he is very much able to walk, talk, have children and do the normal everyday things.

Allah states in His Word that once the heart is cutoff from the light it will not be opened unless He so wills it. So it is Allah Who is in charge and not the head of the prophets – Muhammad (pbh). As illustrious as his soul may be, he is just a human being and does not control the destinies of the people. Can anyone make the blind to see and the deaf to hear in the spiritual sense? The answer is no. Science can make some of those who have physical defects to see and hear physically but never could they open the heart to His Light to make them aware and grateful. Therefore, none but the Creator Lord can make the blind to see and the deaf to hear.

So the dead are those who are buried in their graves in the spiritual sense – deaf, dumb and blind but quite active in the physical realm. That is why the dead can bury the dead and why a person cannot alter his condition of and by himself or with his own self-imaginary power.

Since mankind cannot read the heart and because he has no deep sense of judgment, it would be foolish to pronounce the dead who are doomed to the Fire as being known to us. Only in the case of His declaration, then we can be sure of that person's doom and this can be taken only from His Word or from the privileged word given to a prophet. Otherwise we become engaged in judgment or spiritual terrorism and are false even to our own selves. That of course is what Jesus (pbh) was trying to get across when he asked the people to get the mote out of their own eye first before they went ahead to carelessly criticize others.

One such person who has received the curse was Abu Lahab and his wife also. That can be declared. What cannot be declared are what the pompous arrogant ones declare in their overzealous faith and gigantic ignorance of Hell-Fire and damnation. We cannot comprehend His Wealth of Forgiveness and Mercy nor understand very much how the real world operates. For many, the 'good' and 'bad' people have been classified into various categories yet we should understand that real judgment comes through Him and not by man. Such is the understanding we get from the fact that we do not really understand the dynamics of life as proven by the words of Jesus when he stated, "many that are first shall be last; and the last shall be first." Matt. (19: 30).

All of this should be common sense and the good news is that we were visited by that spirit of truth (Muhammad) who showed the Way clearly. He is that illustrious soul who gave the balanced teachings to the world and the correct sentiments towards life. However, he died but Allah does not!

In Summation

So what we have here is what can be expressed in English after the Fourth Age comes to a close as having the hereafter and after the Sixth Age comes to a close we have the HEREAFTER. It goes something like this for those who understand:

The Fourth Age ends with the return (again) of Jesus the son of Mary (pbh) and the defeat of the Anti-Christ and the destruction of the Gog and Magog and their friends. Billions of people have died in a relatively short time (less than a decade) and society as known before has nearly become extinct. The normal ages of man are over. The Gift of Allah to the faithful begins.

And whosoever obeys Allah and the Messenger (Muhammad) shall be with those whom Allah has blessed – the Prophets, the truthful and the martyrs and the righteous: What excellent companions these are that one may get! This is the real bounty that comes from Allah and Allah's Knowledge suffices to know the Reality. Quran (4: 69-70)

The Fifth Age begins in splendor and glory, as a new heaven and a new Earth are reborn in great spiritual essence. The Satins are locked up. The blessed land of Palestine or Palestine of old shines forth like a bright beacon of Praise to the One God (Allah). Under the Vicegerency of Jesus the Messiah-soul and others, the breaking of crosses and the killing of swine (as defined in my second book) are completed. The new community that emerges is based on the Quran, the Sunnah and the true Shari'ah Law of Prophet Muhammad (pbh) such that man returns to a state of oneness in one community with oneness of intent and worship.

Trade and expansion of society continues and all are engaged in pure worship and harmony and the Earth is blessed and becomes a Garden of repose and worship. The main language in this hereafter will be Arabic and there shall be no hint of stain or defect in this new Earth. The Messiah will perform Hajj and will lead all those in prayer at prayer time except in the case of the presence of that man known as the Mahdi. This will be a sign for the people of the future in that of the precedence of Islam. Souls of the purest and highest rankings will be born in the time period that Jesus will remain active in. This will be followed by lesser souls as time goes on but all of these souls have shown sincerity for the oneness of the brotherhood of prophets and the One God. It will truly be a heaven on Earth. The land of the Arabs will now be covered with plush vegetation and many rivers.

All of the souls that are entering the Earth's domain are

born of women and grow up in the age of Islam. So they are
sent down rank upon rank so that they may be purified.
**What is with you must vanish: What is with Allah will endure.
And We will reward those who practice fortitude according
to their best deeds. Whosoever does righteous deeds, whether
male or female, provided he is a believer, We well certainly
grant him to live a pure life in this world and We will reward
such people (in the Hereafter) according to their best deeds.
Quran (16: 96-97)
Before this We wrote in the Psalms, after the Message (given
to Moses): "My servants the righteous, shall inherit the earth."
Verily in this (Quran) is a message for people who would
(truly) worship Allah. Quran (21: 105-106)
Blessed are the meek: for they shall inherit the earth. Matt. (5:
5)**

No earthly garden can last forever so as much time passes
as Allah has preordained and after the death of Jesus the
kingdom starts to contract (entropy) so much so that Allah
will send a cool breeze and take out every soul that has even
a grain of goodness left in his heart. Now the fuller awareness
of one of the names of the Prophet (pbh) can be put in a more
meaningful context. He is called in Arabic the al-mahi or
in English the obliterator of unbelief. This does not mean to
apply to souls that are destined for Hell-Fire. This applies to
the blessed and exalted kingdom of the Fifth Age through
which many souls will return and go through the period of
life to be purified and then die their second death and then
to be raised up on their best of deeds to be judged.
**Resurrection shall take place so that Allah may reward those
who have believed and done good works... Quran (34: 4)**

Once again this kingdom was promised to the Messiah
and this kingdom of the hereafter (not the Hereafter) will be
one community under the Quran, the Sunnah and Shari'ah
Law which most of the current world rejects. Hence, the term

al-mahi, which denotes the obliteration of falsehood and this term belongs to only Prophet Muhammad (pbh). Did not Jesus (pbh) say that a man must be born again of water and the Spirit? When the children are born into this paradise on Earth, they will be born into Islam. Therefore, in simple terms they will be born again from woman and in the spirit of truth. So once again we have another level of understanding of not only that verse in the New Testament but also of one of the names of Muhammad (pbh) – al-mahi and to the deeper meaning of what that name means. Are you in spirit (incased in a body) members of the human race? Was not Muhammad the spirit of truth a human being? All praise to Allah Who sent His Messenger to all of mankind and to the Jinn also.

One can also understand the special prayer (du'ah) that the Prophet taught the people to say in their earnest appeal to be brought forth (raised up) in very high ranks. This prayer is an appeal by the earnest Muslim that he is given that high place in Paradise that is deserving of the best of peoples. How Allah accomplishes this is dependent on Allah alone and the believer must put his earnest trust and appeal to the One Who is the Sovereign of the universe.

The Sixth Age begins and people are still being born into the world by women giving childbirth. This is the age of Iblis to appear as their 'god' and all manner of evil and deceptions are practiced. Before this age, devil worship could not enter Arabia but now the women from the tribe of Daus will worship their idols and other tribes will worship their lusts as they revert to their true natures.

'A'isha reported: I heard Allah's Messenger (may peace be upon him) as saying: The (system) of night and day would not end until the people have taken to the worship of Lat and 'Uzza. I said: Allah's Messenger, I think when Allah has revealed this verse: "He it is Who has sent His messenger with right guidance, and true religion, so that He may cause it to

prevail upon all religions, though the polytheists are averse (to it)" (9: 33), it implies that (this promise) is going to be fulfilled. Thereupon he (Allah's Apostle) said: It would happen as Allah would like. Then Allah would send the sweet fragrant air by which everyone who has even a mustard grain of faith in Allah would die and those only would survive who would have no goodness in them. And they would revert to the religion of their forefathers. (Muslim)

This reversion is a form of entropy that causes things to decay, degenerate or return to the primordial state. In this case the primordial state, which admittedly is a hard thing to discuss, is the state of the essence of man and his/her willful desire. That is that as a high creation man can submit himself to his Creator Lord willfully (one of the purposes of organized religion) or use his desires for other unscrupulous purposes to fulfill 'primitive' desires or lusts. So, in this case the soul has to face itself and is in the end responsible for its own actions.

The Kabah will be a forgotten thing as to what it was used for and the word 'Allah' will be forgotten such that one man will say to another man that he heard that word (Allah) used by his grandfather but that he doesn't know what it refers to. That is because Allah will take out all wisdom, mercy and forgiveness from those people and let them to be Satan's people.

Then this age will come to a close when a man from Ethiopia will come in order to destroy the Kabah.

Abu Huraira reported Allah's Messenger (may peace be upon him) as saying: The Kabah would be destroyed by an Abyssinian having two small shanks. (Muslim)

When he is leveling it to the ground, a great Shout will issue forth and all the people who have lived will be raised up or regenerated from their tailbone and the Earth will become flat like a table top with no oceans or rivers. The mountains will have become like dust and vanish.

On the Day We shall remove the mountains, and you will see the earth as a leveled flat plain, and We shall gather them all together, nor shall We leave out any one of them. Quran (18: 47)

When the Event Inevitable comes to pass,
Then will no (soul) deny its coming.
(Many) will it bring low; (Many) will it exalt. Quran (56: 1-3)
And those foremost (in faith) will be foremost (in the Hereafter). Quran (56: 10)
The Day that We roll up the heavens like a scroll rolled up for books (completed); even as We produced the first Creation, shall We produce a new one: a promise We have undertaken: truly shall We fulfill it. Quran (21: 104)

This is the Day of Assembly or the Day of Reckoning and the Sun will be brought close, very close, and recreated man will be in terror and will, while standing in awe of the time of rectitude, will start to sweat copiously as if they were surrounded by a super heated furnace. Some that day will have been spared the reckoning. And some that day will have a very light reckoning. And some that day will have only a small reckoning. And some that day will have a heavier reckoning. And some that day will have a much heavier reckoning and so on and so forth as a veritable flood of men and women who were not true to their Lord will have such anxiety and fear in them that the horror will be unbelievable and unbearable but still they cannot move.

On the authority of Abu ad-Darda the holy Prophet said, "Those who have excelled in good works shall enter Paradise without accountability; and those who are following the middle course, shall be subjected to accountability but their accountability shall be light. As for those who have been unjust to themselves, they shall be detained throughout the long period of Resurrection and accountability. Then Allah shall also with His Mercy. And they are the ones who will say,

"Thanks to Allah Who has removed sorrow from us."

According to a tradition, the holy Prophet said, "I declare an oath by Allah, in Whose hand is my life, that the long, horrible Day of Resurrection will be made very short and light for a Believer, as short and light as the time taken in offering an obligatory prayer." (Musnad Ahmad)

The day of truth will have arrived and much of mankind will wish to be a dead and forgotten thing. These are the various times of the soundings of the Trumpets and the passing out of the records of life as has been described in the Quran and hadiths. The Six Ages of man theory does not cover this time but ends only with the destruction of the Kabah as the ending of the Sixth Age.

These are the people who will want to be sent back after the Day of Judgment upon receiving the news of their horrible fate that they will try to escape from. They will claim that they are to be trusted to mend their ways. However, their records will be shown to them and it will be clear to all that they are abject liars loving deception and evil. So, nation after nation and groups after groups will be led into the Fire cursing the nation or group that has preceded it.

Then it will be a single shout and behold, they will begin to see.

They will say, "Ah! Woe to us! This is the Day of Judgment!" Quran (37: 19-20)

If it had been the Lord's Will, they would all have believed – all who are on earth! Will you then compel mankind against their will to believe?

No one can believe without Allah's permission and it is the way of Allah that He throws filth on those who do not use their common sense. Quran (10: 99-100)

But in the end, these very efforts of theirs will become a cause of their regret; then they will be overcome, and the

disbelievers shall be gathered and driven towards Hell: So that Allah may separate the filthy from the pure and gather together every sort of filth and then throw the whole heap into Hell: They are the people who have, in fact, become bankrupt. Quran (8: 36-37)

The Unbelievers will be addressed: "Greater was the aversion of Allah to you than (is) your aversion to yourselves, seeing that you were called to the Faith and you used to refuse."

They will say: "Our Lord! Twice have You made us to die, and twice have You given us life! Now have we recognized our sins: is there any way out (of this)?" Quran (40: 10-11)

Every being in the heavens and the earth will come to Allah as a servant.

He does take an account of all of them and has numbered them (all) exactly.

And every one of them will come to Him singly on the Day of Judgment. Quran (19: 93-95)

Those who reject faith and deny Our Signs will be companions of the Hell-Fire. Quran (5: 10)

This must be so because the unbelieving and ungrateful souls will condemn those who led them astray all to no avail. The previous nations or groups will be in a condemning mode. The strong ones will point out that the weak ones among them had the opportunities to go on the Straight Path but that their own twisted notions of the rejection of the truth overcame their fortitude of entering upon the Way and they but followed nothing but their own selfish desires. In anger the weak ones will curse the strong ones of their community and ask that they will be given double the punishment for leading them astray. It will be announced to all of the condemned ones (the rejecters) that double the punishment will be for both of them! This is due to their willful practices of evil (tahgut) and the fact that others even under torture or worldly hate stood firm in their commitment to the One God and the oneness

of the brotherhood of prophets.
They say: "Shall we really be restored to our former state?"
"What! – when we shall have become rotten bones?"
They say: "It would in that case be a return with loss!"
But it will only need a single shout (compelling cry)
When they will forthwith appear in an open plain. Quran (79:
10-14)
Were We then weary with the first Creation that they should
be in doubt about a new Creation? Quran (50: 15)

QUESTIONS

How many Resurrections are there?

Basically there is only one Resurrection and that would come after the Sixth Age is closed by the heralding of one simple Shout such that the Earth gives up its dead as well as the one's left alive to face the period known as Judgment. Some people have resurrected ideas from the past while some have been known to resurrect the dead but in fact there is only one meaningful Resurrection. All of this is based on purpose, intent and ideal.

So man is put in a test but it is his purpose, intention and the ideal that offers a life raft for security and hopefully not a false sense of security. We can't know the Unseen but we can do one thing and that is to remember Him much. This is not the foolish Eastern practice of rising to perfection by one's own means but it is the awareness of the Creator Lord as the Most Merciful and Most Forgiving.

What is the language of paradise?

The language of the heavenly kingdom on Earth or the Fifth Age will be Arabic. There is a tradition concerning the language of paradise being Arabic. This may be classified as a weak tradition and the wording may not be clear or the understanding of an isolated hadith may not be clear. However, when the understanding is put together piece-by-piece, line upon line and pattern upon pattern, then the understanding becomes clearer. This concept of Arabic being the language of paradise is due to the unified expressions of souls entering the world among other things. This serves as an important lesson for the heaven on Earth or the Messiah kingdom. The unifying book for that age will be the Quran but one must think about this. What good would the Quran do anybody in that age if there were no mountains, rivers or oceans? What good would the Quran do if all the verses about Zakat and trade and so forth and so on do if these things were not to continue? For sure the tax on non-Muslims will cease but the Islamic Shari'ah will be completely in place all over the Earth without exception. That is one of the basis for the Six Ages of man theory.

How is this to be done? One can also ask, "How is it possible that a group of men can find shade under a pomegranate?" Also, souls will be returned in stages and not all at once. The souls will be resurrected, however, at one and the same time.

▓ What is the condition of the Middle East today?

The condition of the Middle East today is as Prophet Muhammad (pbh) described in these perilous times. In short, the star of Israel is on the rise and the star of the Arabs is on the wane. The Wisest of Planners has set the conditions for this. And if you cannot find a test for your life be sure that He will test you under conditions that will cause you to be strained as to your patience and fortitude.

**Do you think that you will enter Paradise without undergoing such trials as were experienced by the believers before you...
Quran (2: 214)**

Let us break the silliness of subterfuge and falsehoods. Do the world's democracies want you to have democracies and peace?

Then the world should take the honest and just road and not one of poorly disguised hypocrisy. So if one wants peace, a just peace with a true democracy, then take the whole of the Middle East and have a vote. Perhaps about 94% of the whole Middle East wants all of the U.N. Resolutions to be followed. It is blatantly hypocritical for only the Arab nations to follow these Resolutions and have Israel exempt from them. What America and Israel have tended to pull off was a diplomatic coup whereby the Muslim, Christian and Jewish Arabs would be held as third class citizens and where Israel and America could control a large population under the false guise of democracy. So take a vote as collective peoples not just by one nation. It is important not to forget the special prayers that the Prophet (pbh) gave when confronting the enemy or when under great stress. Did they forget to call upon in earnest the One Who could help or did they think that they could take time off from their real duties and let others do the work for them? Then which of the favors of your Lord will you deny?

APPENDIX A

... "O my Lord! Increase me in knowledge." Quran (20: 114)

The Quran directs man not only to ponder over the creation but also to seek out knowledge. Seeking out rightful knowledge helps the believer to do several things:

1. His faith becomes stronger as well as his belief in the qualities of Allah.
2. He becomes more acute to the accountability and praise for Allah.
3. He is in a position to earn greater rewards form Allah (Allah Willing).
4. Because he is stronger in awareness, he is less likely to be taken in by 'wonders and deviant notions'.
5. Upon his death all rewards cease except in three cases and one of these is the passing on of knowledge that may help his fellow man to see things more clearly.

However, knowledge in Islam is not willy-nilly or just any kind of knowledge. It must be used correctly and or applied correctly. It is good to have a debate on the purpose of knowledge and the correct application of the same so that certain misunderstandings, when they arise, can be put in a proper perspective. It is with this idea in mind that I turn to Sayyid Maududi's work, *The Meaning of the Quran.*

Knowledge comes in many forms and has many faces. Some of these faces are in the areas of conjecture and suppositions as well as in the science of passing on meanings of the verses of the Quran, the sayings of the Prophet (pbh) and their meanings and comparative religion showing His Oneness and Completeness. Knowledge is akin to truth and so can be used to uplift, enlighten, prepare and explain. Knowledge has to be channeled so that it can do the above things. Therefore, correct knowledge and the use thereof should have its basis in Islam and not philosophical ideology.

What is forbidden is not conjecture as such but excessive conjecture and following every kind of conjecture, and the reason given is that some conjectures are sins. In order to understand this Command we should analyze and see what are the kinds of conjecture and what is the moral position of each. One kind of conjecture is that which is morally approved and laudable, and desirable and praiseworthy from a religious point of view, e.g. a good conjecture in respect to Allah and His Messenger and the believers and those people with whom one comes in common contact daily and concerning whom there may be no rational ground for having an evil conjecture. The second kind of conjecture is that which one cannot do without in practical life, e.g. in a law court a judge has to consider the evidence placed before him and give his decision on the basis of the most probable conjecture, for he cannot have direct knowledge of the facts of the matter, and the opinion that is based on evidence is mostly based on the most probable conjecture and not on certainty...

The third kind of conjecture, which is although a suspicion is permissible in nature and it cannot be regarded as a sin. For instance, if there are clear signs and pointers in the character of a person (or persons), or in his dealings and conduct, on the basis of which he may not deserve to enjoy one's good conjecture, and there are rational grounds for having suspicions against him, the Shari'ah does not demand that one should behave like a simpleton and continue to have a good conjecture about him....

The fourth kind of conjecture which is, in fact, a sin is that one should entertain a suspicion in respect of a person without any ground, or should start with a suspicion in forming an opinion about others, or should entertain a suspicion about the people whose apparent conditions show that they are good and noble...

This analysis makes it plain that conjecture by itself is not anything forbidden; rather in some situations inevitable, in some permissible up to a certain extent and un-permissible beyond it, and in some cases absolutely unlawful. That is why it has not been enjoined that one should refrain from conjecture or suspicion altogether but what is enjoined is that one should refrain from much suspicion. Then, to make the intention of the Command explicit, it has been said that some conjectures are sinful. From this warning it follows automatically that whenever a person is forming an opinion on the basis of conjecture, or is about to take an action, he should examine the case and see whether the conjecture he is entertaining is not a sin, whether the conjecture is really necessary, whether there are sound reasons for the conjecture, and whether the conduct one is adopting on the basis of the conjecture is permissible. Everyone who fears God will certainly take these precautions. To make one's conjecture free and independent of every such care and consideration is the pastime of only those people who are fearless of God and thoughtless of the accountability of the Hereafter. (Taken from The Meaning of the Quran, note # 24 pp. 105-107 vol XIII, Surah AL-Hujurat, verse 12)

Here the Quran is warning man of an important truth. To judge or make an estimate on the basis of conjecture and speculation in the ordinary matters of worldly life may be useful to some extent, although it would be no substitute for knowledge, but it would be disastrous to make estimates and give judgments merely according to one's own conjectures and speculations in a question of such fundamental nature and importance as whether we are, or are not, responsible and accountable to anyone for the deeds and actions of our lifetime, and if we are, to whom we are accountable, when and what shall be the accountability, and what will be the consequences of our success and failure in that accountability. This is not a question on which man may form an estimate merely according to his conjecture and speculation and then stake his entire life-capital on the gamble. For if the conjecture proves to be wrong, it would mean that the man has doomed himself to utter ruin. Furthermore, this question is not at all included among those questions about which one may form a right opinion by the exercise of analogy and conjecture – for conjecture and analogy can work only in those matters which are perceptible for man, whereas this is a question which does not come under perception in any way. Therefore, it is not at all possible that a conjectural and analogical judgment about it may be right and correct. As for the question: What is the right way for man to form an opinion about the matters which are non-perceptible and incomprehensible in nature? This has been answered at many places in the Quran, and from this Surah also the same answer becomes obvious, and it is this: (1) Man himself cannot reach the reality directly; (2) Allah gives the knowledge of the reality through His Prophets; and (3) man can ascertain the truth of that knowledge in this way: He should study deeply the countless signs that are found in the earth and heavens and in his own self, then consider seriously and impartially whether those signs testify to the reality that the Prophet has presented, or to the different ideologies that the other people have presented in this regard. This is the only method of scientific investigation about God and the Hereafter that has been taught in the Quran. Doomed would be the one who discarded this method and followed his own analogies and conjectures. (Taken from The Meaning of the Quran, note # 8 pp. 163-164 vol. XIII, Surah AZ-Zariyat, verse 10)

This verse forbids people to ask useless and unnecessary questions because some people used to put such questions to the Holy Prophet as were of no practical good for mundane affairs nor for spiritual up-lift. For example, once a certain person while sitting in a gathering asked, "Who is my real father?" Likewise, sometimes, some people put unnecessary questions concerning legal matters so as to get these defined, whereas they had been purposely kept undefined for the good of the people. For example, Hajj was made obligatory by a commandment without specifying whether it was to be performed every year or not. When a certain person heard it, he instantly asked, "Has it been made obligatory to perform Hajj every Year?" the Holy Prophet did not make any reply. The man repeated the question, but he again kept quiet. When the man put the question for the third time, the Holy Prophet replied, "Woe to you! If I had said: 'Yes', the performance of Hajj every year would have become obligatory and the people like you would have been unable to perform it and been guilty of disobedience."
The Holy Prophet himself forbade people to ask questions for the sake of asking questions and to probe into things aimlessly. In a Tradition he warned, "The worst offender against the Muslims is the person who asked a question about something

that had not been made unlawful but was made so because of his question." In another Tradition he said, "Allah has prescribed some obligatory duties for you; let not these go unfulfilled and He has made certain things unlawful, do not go near them. He has prescribed certain limits {so} do not transgress them. He has been silent concerning certain things, but not because He has forgotten them; so do not try to probe into such things." (Taken from The Meaning of the Quran note # 116 p. 76 vol.V, Surah AL-Ma'idah, verse 101)

If Islam is under attack, directly or indirectly, from any area including science, politics, from those who are called 'People of the Book', or from malicious individuals, then it is incumbent upon those who can to stand up for what is right and to come to the defense of what is just. There is no question that human beings are only human beings and that human beings are not perfect as they make mistakes in both actions and deeds. However, to shove all things towards Allah and have the express notion that it is only up to Him to extricate the situation is a grave mistake. It is like the Jews of old who said:

They said: "O Moses! We shall never enter it (the holy land) as long as they are in it. Go thou, and your Lord, and fight both of you while we sit here.

He (Moses) said: "O my Lord! I have power only over myself and my brother: So separate us from this rebellious people!" Allah said: "Therefore will the land be out of their reach for forty years: In distraction will they wander through the land: But do not sorrow over these rebellious people. Quran (5: 24-26)

APPENDIX B

The meaning of the term 'wahi' (pronounced wahee) is a difficult concept to understand and includes that which is a life-force's essence.

The lexical meaning of the Arabic word ('wahi') is secret inspiration which is felt only by the one who inspires and the other who is inspired with {it}. The Quran has used this word both for the instinctive inspiration by Allah to His creation in general and for the Revelation towards His Prophets in particular. Allah sends His "wahi" to the heavens with His Command and they begin functioning in accordance with it (XL: 12). He will send this to the Earth with His Command and it will relate the story of all that had happened on and in it. (XCIX: 4-5). He sends wahi to the bee and inspires it with faculties to perform the whole of its wonderful work instinctively (v. 68). The same is true of the bird that learns to fly, the fish that learns to swim, the newly born child that learns to suck milk, etc. etc. Then, it is also wahi with which Allah inspires a human being with a spontaneous idea (XXVIII: 72). The same is the case with all the great discoveries, inventions, works of literature and art, etc., which would not have been possible without the benefit of wahi. As a matter of fact, every human being at one time or the other feels its mental or spiritual influence in the form of an idea or thought or plan or dream, which is confirmed by a subsequent experience to be the right guidance from the unseen wahi.

Then there is the wahi (Revelation) which is the privilege of the Prophets. This form of wahi has its own special features and is quite distinct from all its other forms. The Prophet, who is inspired with it, is fully conscious and has his firm conviction that it is being sent down from Allah. Such a Revelation contains doctrines of creed, commandments, laws, regulations and instructions for the guidance of mankind.

".... Follow the ways made smooth by thy Lord": "work in accordance with the methods which have been taught to thee by Allah's wahi for the smooth running of hive life". It is Allah's wahi (instinctive inspiration) that has taught the bees how to build their wonderful factory with separate combs to rear brood, combs to turn nectar into honey, combs to store food, in short, separate combs to fulfill every aspect of hive life. It is wahi that has taught the bees how to organize themselves into a co-operative society for {a} collective effort to run the "factory" with the queen and thousands of workers to perform a variety of specific tasks. All of these things have been made so smooth for them by wahi that the bees never feel the necessity of ever thinking about it. They have been running smoothly their factory with their collective effort for thousands of years with perfect accuracy. (Taken from The Meaning of the Quran, note # 56-57, pp. 79-80, vol. VI Surah 16 AN-Nahl, verses 68-69)

APPENDIX C

And remember, Jesus, the son of Mary, said: "O Children of Israel! I am the messenger of Allah (sent) to you, confirming the Taurat (which came) before me, and giving glad tidings of a Messenger to come after me, whose name shall be Ahmad." But when he came to them with Clear Signs, they said, "This is evident sorcery!" Quran (61: 6)

This is a very important verse of the Quran, which has been subjected to severe adverse criticism as well as treated with the worst kind of criminal dishonesty by the opponents of Islam...
In this verse the name mentioned of the Holy Prophet (on whom be peace) is Ahmad. Ahmad has two meanings: the one who gives the highest praise to Allah, and the one who is most highly praised by others, or the one who is most worthy of praise among men....
It is also confirmed by history that the sacred name of the Holy

Prophet was not only Muhammad but also Ahmad. Arabic
literature bears evidence that nobody in Arabia had been named
Ahmad before the Holy Prophet...

In order to determine the exact meaning of these passages one
should first know that the language spoken by the Prophet Jesus
and his contemporary Palestinians was a dialect of the Aramaic
language, called Syriac. More than 200 years before the birth of
Jesus when the Seleucides came to power Hebrew had become
extinct in this territory and been replaced by Syriac. Although
under the influence of the Seleucide and then the Roman
empires, Greek also had reached this area, it remained confined
only to that class of the people, who after having access to the
higher government circles, or in order to seek access to them,
had become deeply Hellenized. The common Palestinians used
a particular dialect of Syriac, the accents and pronunciations
and idioms of which were different from the Syriac spoken in
and around Damascus. The common people of the country were
wholly unaware of Greek. So much so that when in A.D. 70 the
Roman General Titus, after taking Jerusalem, addressed the
citizens in Greek, he had to be translated into Syriac. This makes
it evident that whatever the Prophet Jesus spoke to his disciples
must necessarily be only in Syriac.

Secondly, one should know that all the four Gospels contained
in the Bible were written by the Greek-speaking Christians, who
had entered Christianity after the Prophet Jesus. The traditions
of the sayings and acts of the Prophet Jesus reached them
through the Syriac speaking Christians not in the written form
but as oral traditions, and they translated these Syriac traditions
into their own language and incorporated them in their books.
None of the extant Gospels was written before A.D. 70; the
Gospel of St. John was compiled a century after the Prophet
Jesus probably in Ephesus, a city in Asia Minor. Moreover, no
original copy even of these Gospels in Greek, in which these
were originally written, exists....

The third and very vital point is that even after the conquest by the Muslims, for about three centuries, the Palestine Christians retained Syriac, which was not replaced by Arabic until the 9th century A.D. The information that the Muslim scholars of the first three centuries obtained through the Syriac speaking Christians about the Christian traditions, should be more authentic and reliable than the information of those people whom it reached through translation after translation from Syriac into Greek and then from Greek into Latin....

(7) But the decision is not solely dependent on this as to which {Greek word} had St. John actually used in Greek, for in any case that too was a translation of the Prophet Jesus' language, as we have explained above, was Palestinian Syriac. Therefore, the word that he might have used in his good news must necessarily be a Syriac word. Fortunately, we find that original Syriac word in the Life of Muhammad by Ibn Hisham. Along with that we also come to know its synonymous Greek word from the same book. Ibn Hisham, on the authority of Ibn Ishaq, has reproduced the complete translation of 15: 23-27 and 16: 1 of the Gospel according to Yuhannus (Yhanna: John), and has used the Syriac word *Munhamanna* instead of the Greek Paraclete. Then, Ibn Ishaq or Ibn Hisham has explained it thus: The *Munhamanna* in Syriac means Muhammad and in Greek the Paracletus. (Ibn Hisham, vol. 1, p. 248)....

.... Now, if in the translation reproduced by Ibn Ishaq the Syriac word *Munhamanna* has been used, and Ibn Ishaq or Ibn Hisham has explained that its Arabic equivalent is Muhammad and Greek Paracletus; there remains no room for the doubt that the Prophet Jesus had given the good news of the coming of the Holy Prophet himself by name. Along with that it also becomes known that in the Greek Gospel of John the word actually used was Periclytos, which the Christian scholars changed into Paracletus at some later time.

(8) Even an older historical evidence in this connection is

the following tradition from Hadrat 'Abdullah bin Mas'ud: "When the Negus summoned the Emigrants from Makkah to his court and heard the teachings of the Holy Prophet (upon whom be Allah's peace) from Hadrat Ja'far bin Abi Talib, he said: "God bless you and him from whom you have come! I bear witness that he is {the} Messenger of Allah, and he is the same one whose mention we find in the Gospel, and the same one good news about whom had been given by Jesus son of Mary." (Musnad Ahmad). This has been related in Hadith from Hadrat Ja'far himself and also from Hadrat Umm Salamah. This not only proves that in the beginning of the 7th century the Negus knew that the Prophet Jesus had foretold the coming of a Prophet, but also that a clear pointer to "that prophet" existed in the Gospel on the basis of which the Negus did not take long to form the opinion that the Prophet Muhammad (upon whom be Allah's peace and blessings) was that prophet. However, from this tradition one cannot know whether the source of information with the Negus about the good news given by the Prophet Jesus was the same Gospel of St. John, or whether there existed some other means also at that time for this information. (Taken from The Meaning of the Quran note # 8 pp.211-217 vol. XIV Surah AS-Saff, verse 6)

In answer to Maududi's last sentence from above, a very interesting theory presents itself when looking at the connection of the Ethiopian early emigration of the persecuted Muslims. When the pressure and the torment of the pagan Arabs became too intolerable to bear, the little band of struggling Muslims took refuge in Ethiopia where they were given refugee status and protection by a wise king called the Negus. The king recognized in his conversation with the members of this new religion the truth in their claims about Prophet Muhammad. This Negus is looked at as a wise and tolerant man who later changed his religion from Christianity to Islam.

Nature abhors a vacuum and it seems peculiar that out of the blue such a wise man as the Negus showed a great deal of knowledge that he could tell within such a short time about Islam. Perhaps he did have some special knowledge about another prophet to come after Jesus (pbh).

Some Christian researchers have pointed out some very interesting thoughts, which might have a connection to these events. There was a man called Matthew (not the tax collector supposedly responsible for the Gospel of Matthew) but his real name is Mattias. Mattias is credited by a few as being the probable author of the 'Q' document from whence the four canonical Gospels are said to be derived.

Mattias was theoretically the unofficial biographer of Jesus and accompanied him wherever he went while documenting his activities. He was also honored to be the chosen one who was to replace Judas Iscariot thus keeping the number of holy disciples at twelve. Not much is known about this man but it is supposed by some that he ended up in Ethiopia where eventually he died.

It is significant to look at this paper trail and what it might mean. Away from the corrupt influences of Roman and Greek interference the teachings of Jesus, should theoretically at least, be purer in form and last longer than in Rome. However, this would not last forever but what might be supposed is that a purer form of part of the New Testament was in existence in the form of a Gospel by Mattias, if not publicly, then at least available to the Ethiopian king. In fact, it is quite likely that a Gospel by Mattias could still be existent in Ethiopia and hidden somewhere like the Dead Sea Scrolls were hidden just waiting to be recovered.

This fascinating story has not been concretely proven but it does offer exciting possibilities and explains a deeper and more logical connection between the Negus of Ethiopia, the Muslim emigrants and their fortunate choice of Ethiopia, and the wisdom of the Negus based not on impromptu magic but on logic and faith. Although miracles do happen, the concept of planting a seed and seeing what comes of it is more profitable than just waiting around for things to drop from the sky!

Another interesting thing that occurred after the event with the Negus is that one of the heads of Christendom in Egypt gives a fantastic reply to one of the emissaries of the holy Prophet.

Although it has not been mentioned in the Quran as to what it was that the Holy Prophet had forbidden himself, yet the traditionists and commentators have mentioned in this regard two different incidents, which occasioned the revelation of this verse {Q. LXVI: 1}. One of these relates to Hadrat Mariyah Qibtiyyah (Mary, the Copt lady) and the other to his forbidding himself honey.

The incident relating to Hadrat Mariyah is that after concluding the peace treaty of Hudaibiyah one of the letters that the Holy Prophet (upon whom be Allah's peace) sent to the rulers of the adjoining countries was addressed to the Roman patriarch of Alexandria also, whom the Arabs called Muqawqis. When Hadrat Hatib bin Abi Balta'a took this letter to him, he did not embrace Islam but received him well, and in reply wrote: "I know that a Prophet is yet to rise, but I think he will appear in Syria. However, I have treated your messenger with due honor, and am sending two slave-girls to you, who command respect among the Coptics." (Ibn Sa'd). One of those slave-girls was Sirin and the other Mariyah (Mary). On his way back from Egypt Hadrat Hatib presented Islam before both and they believed. When they came before the Holy Prophet (upon whom be peace) he gave Sirin in the ownership of Hadrat Hassan bin Thabit and admitted Hadrat Mariyah into his own household. In Dhil-Hijjah, A.H. 8 she gave birth to the Holy Prophet's son,

Ibrahim. (Al-Isti'ab; Al-Isabah). (Taken from note # 2, p. 377 Surah AT-Tahrim LXVI, verse 1)

This offers proof that knowledge of a new prophet from amongst the Arabs will come in the future as predicted by both the Old and New Testaments. It must be noted that the New Testament was read and studied in the Arabic language, which is closer in style to the language that Jesus used than the Greek or Roman languages.

Another noteworthy thing is that the highly educated students of these records failed to let the knowledge come to the ordinary of the faithful. Hence, they are responsible for covering up the truth or at least not giving full vent to it. Such was the responsibility of the priest class but they caved in to man-made doctrines and cherished them above revealed scripture.

Still another important thing is what the Muqawqis said in his famous letter. He did not declare that it was a possibility or a probability that a new prophet was going to come forth after Jesus. He openly declared that <u>he knew</u> that a prophet was going to appear after Prophet Jesus. There was no doubt in his mind. The only doubt in his mind was the country he was going to appear in. He believes that this mighty happening will come from Syria. This gives us one more piece of good fortune. In order for a prophet to be born in Syria, he would logically be of <u>the Arab line</u> and that thought can be still gleaned from both the Old and New Testaments of today under careful reading of course. So, the result of this goes on to show what it means when Allah declares that people should study the records as they should be studied. That means that people should be actively engaged in their religion while pondering over their scriptures without overly being dependent on a priest class.

APPENDIX D

Although he was only 13 years old at the time of the death of the Prophet (pbh) Abdullah Ibn ʿAbbas became one of the most learned Companions of the Prophet. The Prophet would often sit close to him while praying: "O Lord, make him acquire a deep understanding of the religion of Islam and instruct him in the meaning and interpretation of things." He was also called, by the second caliph of Islam (ʿUmar) "the young man of maturity." Abdullah was acknowledged as one of the most learned of men in Islam.

The huge controversy over the gardens of paradise, for most people at least, would seem to come from the area denoting paradise on this Earth as well as Paradise in the heavens. This issue has a big impact on the validity of the Six Ages of man theory that strongly adheres to the belief in a temporal paradise on Earth and a permanent Paradise in the heavens and the reasons for these things. In this book the paradise on Earth was called the Gift of Allah because man could purify himself if he isn't one of the few who have already done so while being rewarded by doing great deeds.

Then he must die as all men die and of course he must be raised up to a place where death will be no more. The Old Testament and the New Testament do allude to this but in such a sparse way. Also, when studying the records it should be remembered that <u>no time line</u> is put forward so these concepts are difficult to perceive without the study of Islam. As in all things, Allah knows best. Here are some thoughts on this matter from Abdullah Ibn 'Abbas and others.

"Jannat al-ma'wa" literally means "the Jannat (Garden) that is to be an abode." Hadrat Hasan Basri says that this is the same Jannat which the believers and the righteous will be given in the Hereafter, and from this same verse he has argued that that Jannat is in the heavens. Qatadah says that this is the Jannat in which the souls of the martyrs are kept; it does not imply the Jannat that is to be given in the Hereafter. Ibn 'Abbas also says the same but adds that the Jannat to be granted to the believers in the Hereafter is not in the heavens but here on the earth. (Taken from The Meaning of the Quran, part of note # 11 p. 243, vol. XIII Surah AN-Najm, verse 13).

According to the Quran, the earth will take a new shape in the Hereafter:
"The earth will be spread." (LXXXIV: 3).
"The bottoms of the oceans will be split (and the whole water will sink down in the earth.)" (LXXXII): 3).
"The oceans will be filled up." (LXXXI: 6).
"The mountains will be reduced to fine dust and scattered away and there will be left no curve or crease in the earth." (XX: 105-107).
"On that day the earth will be totally changed." (XIV: 48).

"And it will be turned into a garden and given to the pious people to dwell therein for ever." (XXXIX).
This shows that ultimately this earth will be turned into Paradise which will be inherited by the pious and righteous servants of Allah. The whole earth will become one country, and there will be no mountains, oceans, rivers and deserts which today divide it into countless countries and homelands and divide mankind as well into as many tribes, races and classes. Hadrat Ibn 'Abbas and Qatadah have held this same view that Paradise will be established on this very earth. (Taken from The Meaning of the Quran, note # 83 p. 122, vol. VII Surah TaHa, verse 105).

My argument is not that the Earthly paradise will be 'one country' as in one community of believers worshiping with one consent but that this is not the time for the disappearance of the oceans, mountains, deserts and rivers. The reason for my belief is found in several hadiths that describe the land of the Arabs as being marshy with lots of rivers. Also, trade between nations will take place according to Islamic Law.

Imagine the kind of life it would be like when the Quran is The Book and Arabic the Language and the people are reading about travel on the oceans and trade and a whole host of other things that don't exist! That does not make any logical sense. When the oceans and mountains are removed it is time for the Last Day of Judgment and this is when great turmoil and terror will arise such that any woman carrying a child in her womb will drop her load due to such terror and the hair on the heads of the people will turn white out of the fear that they will be experiencing.

In the Six Ages of man theory, the earthly paradise or earthly garden is given to the approved faithful or what can be said as the proven faithful so that they may, Allah Willing, be rendered pure in this life and be fit for the Heavenly Paradise described in so many hadiths by the Prophet (pbh). That then becomes

'The Gift of Allah' to His servants that He so chooses.

The limitation of Islam in this age (Fourth Age) concerns the landmass that was conquered and could be put under the Shari'ah Law. It was 'Umar and Ali that were supposed to have declared the limits to which Islam would spread as in forming a concept of Dar al-Islam. That didn't mean that the Faith of Islam could not be propagated but that the border territories under its control would be limited.

Once again, how can the universal religion for all times, all peoples and all places be limited? Due to the constraints of the Fourth Age, it is on an 'accept or reject' mode for individuals in their freedom of choice but as to the Fifth Age, it becomes that only chosen vehicle and therefore the verse of the Quran (3: 85) takes on a greater meaning:

And whoever seeks a religion other than Islam, it will not be accepted of him, and in the Hereafter he will be one of the losers. Quran (3: 85)

This may mean different things to different people. Above all, it expresses the non-alignment of the Creator Lord as to basic physical things like race, creed, color, wealth and other worldly trappings. Tongue wagging is out while sincere devotion is in. Nationalities and political dominance are out and truth and servitude to that truth while defending it with courage, honesty and honor is in. Barbaric self-willed behavior is out and good character as defined by the prophets is in.

The best part of this bargain is that man is allowed to strive for what turns his soul on. He even gets a chance to go to the 'Throne' itself and ask with trembling heart the chance to serve and be cleaned. When Jesus (pbh) declared that many knock but few enter, he was giving a hint at the kind of person who desires to grab at the golden prize. That prize cannot be given to a defective heart or an arrogant personality nor can it be given to a careless individual who would treat the prize as an insignificant thing.

APPENDIX E

Every nation on Earth has its hypocrites and every Holy Book ever sent down to mankind talks about these villainous creatures that love to try and blow out the Light with their mouths and devise horrendous plots to drive men to confusion and to the pathway of darkness.

Note it well that the hypocrites shall go to the lowest abyss of Hell, and you will not find any helper for them. Quran (IV: 145)

The hypocrites seek to deceive Allah, whereas, in fact, Allah has deceived them. Quran (IV: 142)

Ever since the first prophet came to mankind, the hypocrites have been attached to man as a test. Towards the end of this current age (Fourth Age) there will be the 'great war' in which one third of the Muslim army will turn on their backs and run away and will never be forgiven by Allah. This hadith given by the Prophet (pbh) shows the flimsiness of faith professed by these so-called Muslims at that time. Islam does not hide the truth nor does it whitewash the realities.

The rejecters of Al-Islam are called disbelievers because they can feel free (or so they think) to accept some things from their Creator Lord and to reject other things like certain commandments or prophets. Therefore, they are paltry believers ranked among the disbelievers but of a higher rank than those who totally reject faith.

The hypocrites, on the other hand, represent wolves in sheep's clothing and are very dangerous because the knife in the back has been expressed with more loathing than a mere frontal assault. Maududi's explanation of hypocrisy covers many footnotes in his Tafsir with various examples and shades of meaning. One very succinct piece follows.

The hypocrites of every age enjoy and have always enjoyed all the benefits conferred by Islam, by professing it with their tongue, and nominally joining the Muslim Community. At the same time they enjoy all the benefits they can derive from the disbelievers, by mixing with them and assuring them, "We are not bigoted Muslims though we are nominally connected with them. We are akin to you in culture, in thoughts and in the way of life and our interest and loyalties are the same as yours. Therefore, you should rest assured that we will side with you in the conflict between Islam and Kufur." (Taken from The Meaning of the Quran, note # 171 p. 176 vol. II, Surah AN-Nisa, verse 141).

Most of the world is locked in a system of disbelief to varying degrees from the very slight to the deep expressions of arrogance. It takes a courageous person to let go of culture, tradition, old wives tales and most importantly pride and arrogance to develop an attitude of truth. It can't be done in people who have a crooked diseased heart or who are impudent schemers.

For example, the term 'Israel', the name given to Jacob (pbh), means servant of Allah. A close look at the real truth of Israel (the physical country called Israel today) will show people that this term would be used for those that followed the Divine Guidance and not for common every day use as some people would like to have one believe. Hence, the term of 'holy Israel' I used in my first book did not apply to the physical Israel found in today's world. Maududi's Tafsir, in a balanced way, brings to light valuable insights as to the Jewish temperament. What can one be afraid of when the truth is uncovered?

An Islamic tradition of the Prophet says that one will be in the Hereafter with the one that he loves. This is great news for the true believers but utter degradation and horror for the criminals and because all souls took a covenant before coming to Earth the claim of ignorance will not shield anyone from the Hell-Fire.

However, help comes in many ways and the best of ways is in understanding Islam and not uncontrolled violence. So help comes from the understanding of fortitude and endurance. How will Jesus and his companions deal with the Gog and Magog? They won't use weapons. They won't be engaged in useless discussions. They will use prayer and special du'ah in their trust in the One True God by calling on Him with sincerity and earnestness. These are the weapons that make tanks, planes and bombs look like nothingness in today's world as Allah Wills. These are the weapons of terror of a higher quality than the gun. It is not that the sword doesn't have its place because it does but the Word will always be mightier than the sword.

The world is after all the world and the worldly don't give a hoot about the awareness of God's Divine Laws. What the worldly people want is their vested self-interest to be looked after and petted. This is why the ignorant (uninformed or uneducated) peoples are often dazzled by speeches and promises luring them into false hopes while manipulating their emotional common sense. Hence, as the complete records point out, billions of people must die to prove a point. The point is that people are indeed accountable for their actions and belief systems and the one who refuses to humble himself to the Divine Will and ask for His Mercy, Forgiveness and Guidance will not prosper on the Last Day of Judgment. On the contrary, they will be exposed as transgressors and be thrown into the Fire.

The recent war between Hezbollah and Israel gives one pause to think. Wisdom has been known to accompany man as well as folly. Common sense should dictate the future course of balanced thinking. Israel has a right as a country, until it foregoes that right, to live in peace within secure borders and so doesn't every other country. Israel has the right to maintain arms that are sufficiently strong enough to use offensively and defensively to enable it to protect itself and so doesn't every other country. It is pure hypocrisy to try and tell other countries that they must disarm or be unprepared for any aggression because it might hurt the feelings of America or Israel or that it is a bad policy to protect and defend oneself because America and Israel are trustworthy while America and Israel are allowed to arm themselves.

The katoucha rockets fired into Israel were so primitive and untrustworthy that they could be called giant firecrackers without much of any guidance system. However, in ten or fifteen years advanced technology can produce rockets that will have a range of nearly 200 miles and carry huge payloads of dangerous explosives. They could conceivably also carry payloads of other things too chilling to think about. Not only that, but also these rockets could presumably be, with new technology, smaller in size than a katoucha rocket and able to be assembled in stages undercover such that they could be hidden from detection.

This idea does not bode well for narrow-minded people who so far have managed to sink United Nations' Resolutions when fair play was called for. However the case may have been in the past, the future immediate peace for the Middle East should not be based on a too long failed American/Israeli policy.

Why should other neighborly countries help Israel and police their own citizens when the Arabs have been treated as third-rate people and lied to even before the disgraceful so-called Balfour Agreement signed in 1917? Without going into politics or past lunacies, the next war could be really something very destructive. So why not nip it democratically in the bud? That is to say, why not show some practical common sense and flush the hypocrites out into the open. The first independent state that fails to comply with the United Nations' Resolutions should be treated as a rouge state and boycotted from any trade and have all of its assets frozen. This would ensure that any tough-talking country or any hypocritical manipulator would be punished and left to rethink its position of intransigence.

The Arab countries as well as all other countries have agreed that all established countries belonging to the United Nations have the right to exist. Israel being a charter member of the U.N. does have the right to exist. It also has the right and the duty imposed on it to obey all of the international treaties and U.N. Resolutions. What goes on outside its borders is not its business as long as there is neither subversion aimed at it nor any subversion aimed against its neighbors. Therefore, Israel must have secure and well-defined borders instead of amorphous ones that seem to expand as time goes by.

They have no right to guarantee peoples' rights outside these well-defined borders as in interfering in another state. They have the right to complain about the treatment of its citizens in another country and pointing out any discrepancies of human treatment by showing the evidence in front of the U. N. but they have no right to take any action detrimental to peace outside its defined and concrete borders because its citizens residing in another country are free to leave that country and reside in Israel.

Therefore, in seeing that the future of peace looks bleak in the Middle East, the U.N. needs to be given greater powers to ensure that full and biting sanctions would be put in place regarding any renegade state (a state not abiding by its agreements) and any mischievous helpers of that state. Full adherence to this rule and a strong will to enforce it, sends out a notice that hanky-panky politics and double crossing hypocrisy will not be tolerated or encouraged.

The U.N. is not put in place to demand that certain philosophies, races or creeds run any country. It is put in place to see that commitments are honored and not abrogated by mere whims of paranoiac people or great soothsayers of doom. In this way, all legitimate countries are put on one level playing field while encouraging each other to play by the rules and discouraging each other from acting contrary to the good intent of their neighbors.

It has been rumored that Israel has several atomic bombs stockpiled somewhere in their country. That is fine with me. They can have a thousand of those bombs. The point is that no one should meddle in their internal affairs. Likewise, if any of their neighbors wish to import thousands of missiles capable of delivering one thousand pound bombs and capable of reaching targets as far away as 400 miles then that is their business and Israel has no right to meddle in another country's internal affairs. O Israel, other countries are not your servants nor are they your surrogates to manipulate at will.

Because a democratic regime is supposed to be responsible, let it be free to be responsible. It is unacceptable in this modern age to have one country lord it over others.

For over thirty years the big issue has not been Israel's right to exist no matter what dodge they have given to the world body and what America has preached to the world. The real facts are quite different and show that all the countries surrounding Israel would accept them and their right to exist. However, the real reason for all the brouhaha in the Middle East is that Israel has no real fixed borders and therefore could conceivably expand to take in all of Lebanon, Syria, Egypt and parts of Saudi Arabia etc by one form or another. This is based on the land grab policies as it slowly expanded from a set and established territory since 1948.

Non-native Israelis within an existing country that had no vote on the matter established the territory of Israel. That shoots democracy out the window right there. However, letting bygones be bygones the Israelis cannot get any respect from its neighbors due to its democratic policies of treating the native inhabitants as bloodthirsty 'sand niggers'. Their so-called two-state solution will be to put a few tightly controlled cities appropriately called reservations but window-dressed up as a separate sovereign nation called a Palestinian State. And this hypocritical dodge is put down as 'democracy in action'.

All of Israel's neighbors are ready to guarantee Israel's right to exist based on a fair and just implementation of the U.N. Resolutions. So-called extremist countries or Israel's non-direct neighbors may have another agenda but they are not Israel's neighbors and they would be dealt with swiftly and painfully as a dangerous rouge state by all the Middle Eastern countries as well as the world at large. Hence, the wind has just been knocked out of Israel's sails and they and their friends are shown up as the real impediments to peace. As one man once said, "Come on gentlemen, you can't end up having it both ways. So stop bringing to the table two faces."

The reign that they are given has deluded them. They think that since they are not being punished in spite of their denial, they must be on the right path whereas they are following the path of ruin.

According to Maududi, "What distinguishes a Muslim from a non-Muslim is that the latter claims absolute freedom but the former considers himself to be the servant of Allah and uses only that amount of freedom which Islam allows him. The non-Muslims judge all matters in accordance with the rules and regulations made by themselves and do not believe that they stand in need of Divine Guidance. In contrast to this, Muslims first of all, turn to Allah and His Messenger for guidance about everything and abide by their decision."

A saying from Caliph 'Umar – "The best way of punishing the one who does not fear Allah in his dealings with you is that you should fear Allah in your dealings with Him (Allah)." This is a good expression from Caliph 'Umar. Students of comparative religion might appreciate this saying as it enlightens one on the correct meaning of the expression of Jesus the son of Mary (pbh) when he declared to his followers concerning the 'turning of one's cheek'. This shows the superiority of a strong believer over a weak one and it also shows how one, with Allah's permission, is putting 'fire' on the head of the disbeliever.

It can be argued from this that 'we' will suffer much but cannot win. The idea in this world is not to win the intangible stuff and to live forever but to do several things. The right approach is to do the honorable things in dignity and in justice. Of course that takes patience, time, effort and the willingness to suffer. Why must people suffer so much? We haven't entered paradise yet and this world is a breeding ground for the careless, the disbeliever and those who are moving in channels of making up their minds as to what side of the fence they want to land on. These things are not a 'quickie' proposition whereby man has instant access to upright behavior.

The Western philosophy of 'peace on Earth and good will to mankind' is according to the Quran an illusion. It won't be an illusion in an earthly paradise to come but in the ordinary ages of mankind, Earth is seen as a testing ground to the submission of truth, it is just a testing ground. So, man must ever be engaged in toiling upwards to establish his record of deserving this paradise.

The disbeliever is likened to a rock that is hard. Hard of hearing, seeing and understanding and believing that he will get his way and enjoy superior conditions for the totality of his life. Like a rock that has no brain or is incapable of thought, he does not understand the peril he is creating for himself. He can neither create his past, nor his present or future. If the rock remains in its place, it will be moved. Even the mountains, as scientists know so well, have been taken down and removed. It may have taken millions of years but they have been moved.

The much better example is found in the New Testament whereby Jesus (pbh) describes the parable of the sower that explains all things in this connection and is a great example of actual life on Earth, Matt. (13: 3-32). But minds grow careless and hearts grow cold and so man tries to accomplish what devious things he can try and get away with and while being reminded of wisdom or the brotherhood of prophets, he grows angry and contemptuous and his heart decays and receives no light. Hence, the peril and the judgment in that he had no real confidence in his Creator Lord.

CONCLUSION

The Quran lives! History, science, the past, present and future are all locked up in that Book. Mankind in general says, "How can that be?" It can be because of the One Who made it so. Far, far above is He, above the common thoughts of mankind.

So, if man wishes to understand, he should divest himself of his ego but that doesn't seem possible right now and so the world continues to turn much like it has always turned and people plot and plan and life goes on and on. It may look like an endless road but that is not the case. Definitely that will not be the case!

If man could only be like a 'god' and control his destiny on Earth, this planet would be even in a much worse state than it is now. Honorable vicegerent or pathetic, obtuse human being or all things in between is what we have become. This, of course, does not apply to the pompous ones as they stride through life heedless of any direction but what their egos lead them to.

Some have taken a man and lifted him up to the level of Godhood while others dance around in some magical assurance that God is only for their pleasure and manipulation while others move around like simple cattle – thinking little and knowing less. The world keeps turning around.

The simple truth is that they cannot control you. They cannot control your thoughts, your actions, your beliefs, and your desires to know. They can threaten and they can try and watch but they can never get into one's heart. So it is imperative to think for one's self, weigh the evidence and choose the system of belief that makes the greatest contribution to truth. Choosing a system that feels good is like asking a small child what he would like for dinner – vegetables or candy? So, people have to choose with care.

What the rest of the 'advanced' world may think of pre-human evolution does not affect the truth whatsoever. This book showed that those theorists are operating in the dark although fairness demands that their findings be recognized. Recognized yes, but not embellished such that simple deductions turn into a horror story for the futuristic development of the human race, which in their case has turned out to be simply a nonexistent fantasy.

Comparative religion has two parts to it: the coordination of truth and the composition of lies. The first part shows the harmony and unity of the one brotherhood of prophets while the second part shows up the twisted attempts of the unscrupulous to become deviant. Both sets have been promised their just reward and they will get it!

If man's intended goal were to seek out mysteries, then for the most part he would be numbered among the failures. Man's apparent goal is to live and develop according to the right-guided practices set forth by Divine Wisdom and given by Divine Revelation minus cultural hoopla or philosophical/political doubletalk. A human being stripped from his clothes, country, culture and physical body represents the real human being. Therefore, it is only logical that a misuse of the above characteristics causes us trouble in trying to find our real selves. Going back to Divine Revelation is like going back to one's foundation. The stronger the foundation, the stronger the edifice! The Quran lives!

Religious leaders and perhaps most important of all, the common man, ask, "Is it true that in Islam Jesus the Christ is the king of Israel?" Of course is the answer. Who did you think it was? After reading the four books written by this author the idea should be clear enough. The idea that Jesus the Christ (pbh) is the true Messiah means that he is the king of Israel. Naturally, if man can forget the overemphasis on the state of his physical dependency on culture, creed, nationality and skin color, he should be able to understand the simple truth that Israel as a country of people has nothing to do with the spiritual Israel.

The spiritual Israel is nothing more and nothing less than the body of true believers who have stood fast by their One Creator Lord and who have stood fast to the one body of prophets. They are they who welcome the paradise on Earth to be governed by the religion of Islam, the Quran and the Sunnah of Prophet Muhammad (pbh). These things are what the records show and no matter how one tries, one cannot destroy the truth of this matter. This is what the Six Ages of man theory states and that is a plain, simple fact.

For too long people (mostly the so-called experts) have danced around the so-called love that they supposedly show for the Messiah and could not address the issue of 'others' as it pertained to Paradise. When some of these people spoke about Paradise it was usually in the vain of rewarding themselves and their own kind that distinction even though their own records forbade it. While the common people, who found no time to read or understand the deeper meanings behind 'religion' just kept meandering through their daily life trusting in the 'wisdom' of those men. Well, it happens.

Now, as we approach some interesting times before the second coming of Jesus (pbh), it would be wise to consider investigating the truth about Islam – and this is a good thing and not just for others. For in this world there are many who go under the name of Islam but have very little idea as to what it is about. It would be helpful to get a copy of Sayyid Maududi's *The Meaning of the Quran* especially for non-Arabic speakers and investigate not only the Quran but also the life and times of Prophet Muhammad as it unfolded. Like in all things that is up to the individual and so he or she must use his or her God-given mind and try to use it well. The Quran lives!

For example, if one thinks he knows Jesus (pbh) what does one make of this mysterious quote from the New Testament?

But whosoever shall deny me before men, him will I also deny before my Father which is in heaven.

Think not that I am come to send peace on earth: <u>I came not to send peace</u>, but a sword.

For I am come to set a man at variance against his father, and the daughter against her mother, and the daughter in law against her mother in law. Matt. (10: 33-35)

It seems so uncharacteristic of that holy prophet! It seems so but it is not! For he is not declaring anything but the truth, yet it would appear otherwise. What he is declaring is that his teachings will be taken in many different ways such that strife will arise between those who seek and follow the truth as against those who will take their own desires as a god and therefore cause hurt, disruption and discord to the unity of his teachings and the principles he laid down. How can we know this for sure? Simply by going into the Quran and seeing for ourselves how the individual operates. The individual, in the case of non-surrendering to Divine Truth, operates by his or her own desires and sees the world and others in a different light than what really is.

Say: "Everyone acts according to <u>his own disposition</u>: but your Lord knows best who it is that is best guided on the Way." Quran (17: 84)

The above Quranic verse is saying that one brings his or her feelings or interpretations with oneself and it is these interpretations that, if one should listen to them, gives one the interpretation to act on things in according to one's desires. Therefore, if the self is corrupt, the belief will be corrupt also because of the self's desire to follow vain practices, thoughts, or actions. Hence, the warning given by Jesus to make the fruit good and therefore the tree good or make it corrupt because the tree is known by its fruit. Matt. (12: 33)

Indeed the Quran gives forth its answers and helps people to understand life but only if people can free themselves from the chains of ignorance and pride. A person can read a book from cover to cover but if the heart is not in it, the mind will not be able to comprehend much. Truly, the Quran lives!